F 5

Student Solutions Manual for

Introduction to Probability and Its Applications

Second Edition

Richard L. Scheaffer
University of Florida

Prepared by
Michael Allen
Chris Franklin

Duxbury Press
An Imprint of Wadsworth Publishing Company
I(T)P™ An International Thomson Publishing Company

Belmont • Albany • Bonn • Boston • Cincinnati • Detroit • London • Madrid • Melbourne
Mexico City • New York • Paris • San Francisco • Singapore • Tokyo • Toronto • Washington

Printed in the United States of America
6 7 8 9 10—04

For more information, contact Wadsworth Publishing Company.

Wadsworth Publishing Company
10 Davis Drive
Belmont, California 94002
USA

International Thomson Editores
Campos Eliseos 385, Piso 7
Col. Polanco
11560 México D.F. México

International Thomson Publishing Europe
Berkshire House 168-173
High Holborn
London, WC1V 7AA
England

International Thomson Publishing GmbH
Königswinterer Strasse 418
53227 Bonn
Germany

Thomas Nelson Australia
102 Dodds Street
South Melbourne 3205
Victoria, Australia

International Thomson Publishing Asia
221 Henderson Road
#05-10 Henderson Building
Singapore 0315

Nelson Canada
1120 Birchmount Road
Scarborough, Ontario
Canada M1K 5G4

International Thomson Publishing Japan
Hirakawacho Kyowa Building, 3F
2-2-1 Hirakawacho
Chiyoda-ku, Tokyo 102
Japan

Printer: Malloy Lithographing, Inc.

ISBN 0-534-23791-6

CONTENTS

CHAPTER 2
PROBABILITY

2.1 There are 10 elements in P, 5 in M, and 13 in neither P nor M. Since there are only 25 elements in all, $P \cap M$ must contain $28 - 25 = 3$ elements. This setting is described by the following Venn diagram.

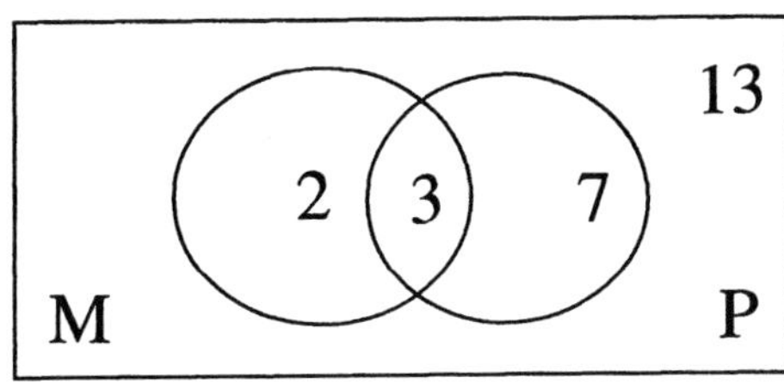

a. PM contains 3 elements.

b. $\overline{P}\,\overline{M}$ contains 13 elements.

c. $P\overline{M}$ contains 7 elements.

d. $(P\overline{M} \cup \overline{P}M)$ contains $2 + 7 = 9$ elements.

2.3 Verification of $A\,(B \cup C) = AB \cup AC$:

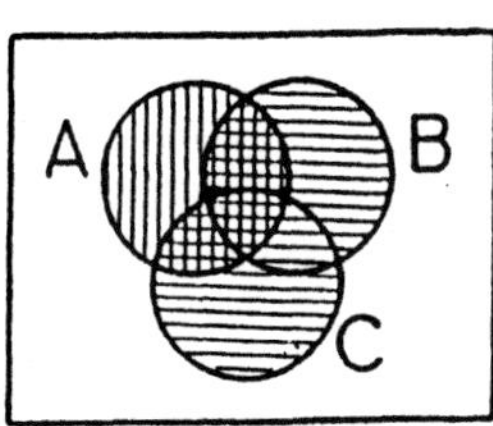

for A, for $B \cup C$

Crosshatched region is

$A\,(B \cup C)$

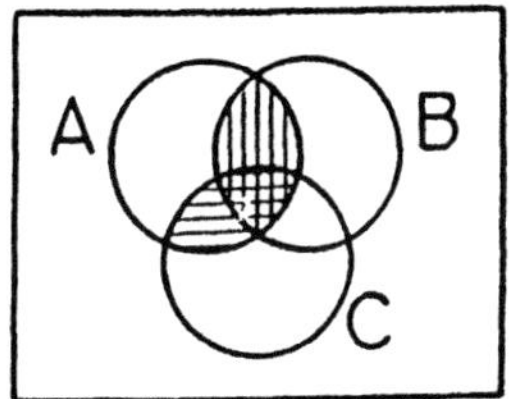

for A, 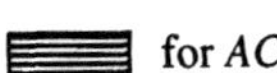 for AC

Region shaded at least once is

$AB \cup AC$

Verification of $A \cup (BC) = (A \cup B)(A \cup C)$:

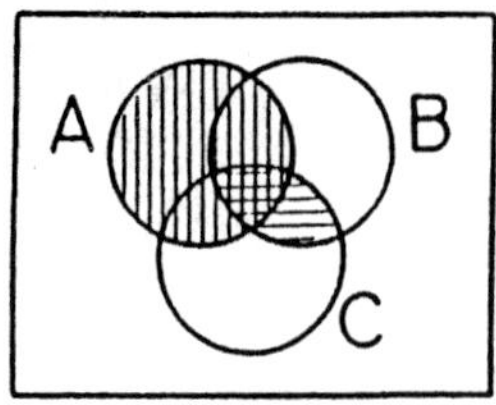

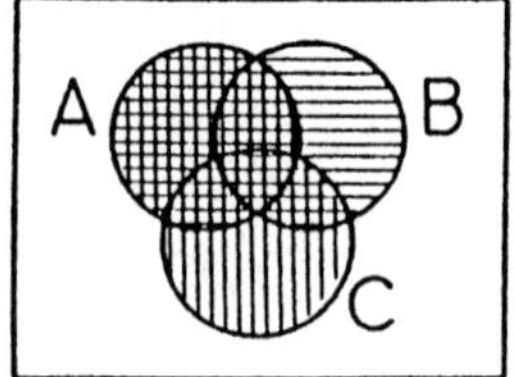

▥ for A, ▤ for BC

Region shaded at least once is

$A \cup (BC)$

▥ for $A \cup C$, ▤ for $A \cup B$

Crosshatched region is

$(A \cup B)(A \cup C)$

2.5 **a.** L = left turn, R = right turn, S = straight

b. $P(L) = P(R) = P(S) = \frac{1}{3}$

c. $P(\text{turn}) = P(L \cup R) = P(L) + P(R) = \frac{1}{3} + \frac{1}{3} = \frac{2}{3}$

2.7 **a.** $\frac{1}{3}$

b. $\frac{1}{3} + \frac{1}{15} = \frac{6}{15}$

c. $\frac{1}{3} + \frac{1}{16} = \frac{19}{48}$

d. $\frac{1}{3} + \frac{1}{3} = \frac{2}{3}$

2.9 Let B denote the event that the assembly has a bushing defect and S the event of a shaft defect.

a. $P(B) = 0.06 + 0.02 = 0.08$

b. $P(B \cup S) = 0.06 + 0.02 + 0.08 = 0.16$

c. $P(B\bar{S} \cup S\bar{B}) = 0.06 + 0.08 = 0.14$

d. $P(\overline{B \cup S}) = 1 - 0.16 = 0.84$

2.11 Denote an outcome as an ordered pair of letters, where the first letter of the pair signifies the outcome for the first vehicle (and may assume the values S = straight, L = left turn, R = right turn) and the second letter signifies the outcome for the second vehicle (and may similarly assume the values S, L, or R).

a.

Vehicle I	Vehicle II	Outcome
Straight	Straight	SS
Straight	Right	SR
Straight	Left	SL
Right	Straight	RS
Right	Right	RR
Right	Left	RL
Left	Straight	LS
Left	Right	LR
Left	Left	LL

b. P (at least one turns left) = $P(SL) + P(RL) + P(LS) + P(LR) + P(LL)$

$$= \frac{1}{9} + \frac{1}{9} + \frac{1}{9} + \frac{1}{9} + \frac{1}{9} = \frac{5}{9}$$

c. $P(\text{at most, one vehicle turns}) = P(SS) + P(SR) + P(SL) + P(RS) + P(LS)$

$$= \frac{1}{9} + \frac{1}{9} + \frac{1}{9} + \frac{1}{9} + \frac{1}{9} = \frac{5}{9}$$

2.13 Denote an outcome as an ordered pair of Roman numerals, where the first Roman numeral of the pair signifies the firm (either I, II, or III) that receives the first contract and the second signifies the firm that receives the second contract.

a. The simple events are (I, I), (I, II), (I, III), (II, I), (II, II), (II, III), (III, I), (III, II), (III, III).

b. P(both contracts go the same firm) = P [(I, I)] + P [(II, II)] + P [(III, III)]

$$= \frac{1}{9} + \frac{1}{9} + \frac{1}{9} = \frac{1}{3}$$

c. P(firm I gets at least one contract) = P [(I, I)] + P[(I, II)] + P[(I, III)] + P[(II, I)] + P[(III, I)]

$$= \frac{1}{9} + \frac{1}{9} + \frac{1}{9} + \frac{1}{9} + \frac{1}{9} = \frac{5}{9}$$

2.15 **a.** $P_2^7 = 7(6) = 42$

b. $\binom{7}{2} = \frac{7(6)}{2(1)} = 21$

2.17 $P_4^{10} = 10 \cdot 9 \cdot 8 \cdot 7 = 5{,}040$

2.19
$$\binom{n-1}{r-1} + \binom{n-1}{r} = \frac{(n-1)!}{(r-1)!\,(n-r)!} + \frac{(n-1)!}{r!\,(n-1-r)!}$$
$$= \frac{r(n-1)!}{r(r-1)!\,(n-r)!} + \frac{(n-r)\,(n-1)!}{(n-r)\,r!\,(n-1-r)!}$$
$$= \frac{r(n-1)! + (n-r)\,(n-1)!}{r!\,(n-r)!}$$
$$= \frac{n!}{r!\,(n-r)!} = \binom{n}{r}$$

2.21 **a.**
$$P\left(\begin{array}{c}\text{one blue, one white,}\\ \text{and one green ordered}\end{array}\right) = \frac{\left(\begin{array}{c}\text{number of ways one}\\ \text{blue, one white, and}\\ \text{one green can be ordered}\end{array}\right)}{\left(\begin{array}{c}\text{total number of ways}\\ \text{a sample of three}\\ \text{can be chosen}\end{array}\right)}$$
$$= \frac{3!}{4^3} = \frac{6}{64} = \frac{3}{32} = 0.09375$$

b.
$$P\left(\begin{array}{c}\text{two blues}\\ \text{are ordered}\end{array}\right)$$
$$= \frac{\left(\begin{array}{c}\text{number of ways of}\\ \text{ordering two blues}\\ \text{out of an order of three}\end{array}\right)\left(\begin{array}{c}\text{number of ways (colors)}\\ \text{the remaining car can}\\ \text{be ordered}\end{array}\right)}{\left(\begin{array}{c}\text{total number of ways}\\ \text{a sample of three}\\ \text{can be chosen}\end{array}\right)}$$
$$= \frac{\binom{3}{2}\binom{3}{1}}{4^3} = \frac{9}{64} = 0.140625$$

c. Note that if no blacks are chosen, then each order may be any of the three remaining colors, and this may be done in $3^3 = 27$ ways. Thus

$$P(\text{at least one black}) = 1 - P(\text{none are black}) = 1 - \frac{3^3}{4^3} = 1 - \frac{27}{64} = \frac{37}{64} = 0.578125.$$

d. Multiplying the answer to (b) by the number of ways to choose a duplicated color, we have

$$\frac{\binom{4}{1}\binom{3}{2}\binom{3}{1}}{4^3} = \frac{36}{64} = 0.5625.$$

2.23 **a.** $P_4^4 = 4! = 24$

b.
$$\text{Required probability} = \frac{\begin{pmatrix}\text{number of orderings}\\ \text{in event of interest}\end{pmatrix}\begin{pmatrix}\text{number of ways the}\\ \text{remaining three operations}\\ \text{may be performed}\end{pmatrix}}{\begin{pmatrix}\text{total number of ways}\\ \text{the operation}\\ \text{can be performed}\end{pmatrix}}$$

$$= \frac{2 \cdot 3!}{4!} = \frac{1}{2}$$

2.25 **a.** $P(2k \text{ runs}) = \dfrac{2\binom{m-1}{k-1}\binom{n-1}{k-1}}{\binom{m+n}{n}}$ where m is the number of defectives and n is the number of nondefectives. In this case,

$$P(4 \text{ runs}) = P(2(2) \text{ runs}) = \frac{2\binom{3-1}{2-1}\binom{7-1}{2-1}}{\binom{3+7}{7}} = \frac{2\binom{2}{1}\binom{6}{1}}{\binom{10}{7}} = \frac{2(2)(6)}{120}$$

$$= \frac{24}{120} = \frac{1}{5}. \text{ No}$$

b. $P(\text{2k+1 runs}) = \dfrac{\binom{m-1}{k}\binom{n-1}{k-1}+\binom{m-1}{k-1}\binom{n-1}{k}}{\binom{m+n}{n}}$ where m is the number of defectives and k is the number of nondefectives. In this case,

$$P(3 \text{ runs}) = P(2(1)+1 \text{ runs}) = \frac{\binom{3-1}{1}\binom{7-1}{1-1}+\binom{3-1}{1-1}\binom{7-1}{1}}{\binom{3+7}{7}}$$

$$= \frac{\binom{2}{1}\binom{6}{0}+\binom{2}{0}\binom{6}{1}}{\binom{10}{7}} = \frac{(2)\,(1)+(1)\,(6)}{120}$$

$$= \frac{8}{120} = \frac{2}{30}.$$ Perhaps

c. $$P(2 \text{ runs}) = P(2(1) \text{ runs}) = \frac{\binom{3-1}{1-1}\binom{7-1}{1-1}}{\binom{3+7}{7}} = \frac{2\binom{2}{0}\binom{6}{0}}{\binom{10}{7}}$$

$$= \frac{2\,(1)\,(1)}{120} = \frac{2}{120} = \frac{1}{60}.$$ Yes

2.27 The sample space is $\{SS, SR, SL, RS, RR, RL, LS, LR, LL\}$. Let A be the event of at least one vehicle turning left and B be the event that at least one vehicle turns. Assuming equally likely outcomes, we have

$$P(A|B) = \frac{P(AB)}{P(B)} = \frac{P(A)}{P(B)} = \frac{P(SL \cup RL \cup LS \cup LR \cup LL)}{1-P(SS)} = \frac{5/9}{8/9} = \frac{5}{8}.$$

2.29 **a.** $\dfrac{46,263}{92,911} = 0.4979$

b. Let V = motor vehicle accident, M = male.

$$P(V|M) = \frac{P(VM)}{P(M)} = \frac{32,949}{64,053} = 0.5144$$

c. Let Y = the event of $(15 \le \text{age} \le 24)$. Then

$$P(V|Y) = \frac{P(VY)}{P(Y)} = \frac{14,738}{19,801} = 0.7443$$

d. Let F = event of accidental death by falling, and E = event of age ≥ 75. Then

$$P(F|E) = \frac{P(FE)}{P(E)} = \frac{7,067}{16,065} = 0.4399$$

e. $P(M) = \dfrac{64,053}{92,911} = 0.6894$

2.31 **a.** 0.10

b. 0.03

c. 0.06

d. Let M = mobile home fire, S = fire caused by smoking. Then

$$P(M|S) = \frac{P(MS)}{P(S)} = \frac{P(S|M)P(M)}{P(S)} = \frac{0.06\,(0.03)}{0.10} = 0.018$$

2.33 **a.** $\dfrac{\binom{18}{2}}{\binom{20}{2}} = \dfrac{18 \cdot 17}{20 \cdot 19} = 0.8053$

b. P(at least one is nondefective) $= 1 - P$(both are defective)

$$= 1 - \frac{\binom{2}{2}}{\binom{20}{2}} = 1 - \frac{2}{20 \cdot 19} = \frac{378}{380} = 0.9947$$

c. $P(A|B) = \dfrac{P(AB)}{P(B)} = \dfrac{P(A)}{P(B)} = \dfrac{0.8053}{0.9947} = 0.8096$

2.35 **a.** Let A_i = event that i^{th} resistor has resistance between 9.5 and 10.5 ohms, i = 1, 2. Then

$P(A_i) = 1 - 0.05 - 0.10 = 0.85$, and

P(both have actual values between 9.5 and 10.5)

$$= P(A_1A_2) = P(A_1)P(A_2) = (0.85)\,(0.85) = 0.7225.$$

b. Let E_i = event that i^{th} resistor has resistance in excess of 10.5 ohms, i = 1, 2. Then

P(at least one has an actual value greater than 10.5)

$$= P(E_1E_2 \cup E_1\bar{E}_2 \cup \bar{E}_1E_2) = P(E_1)P(E_2) + P(E_1)P(\bar{E}_2) + P(\bar{E}_1)P(E_2)$$

$= (0.1)(0.1) + 2(0.1)(0.9) = 0.19$ *or*

P(at least one has an actual value greater than 10.5)

$$= 1 - P(\bar{E}_1\bar{E}_2) = 1 - (0.9)^2 = 0.19.$$

2.37 Let C_i = event that relay i closes properly, $i = 1, 2$. Then

P(current flows in series system) = P(both relays are closed)

$$= P(C_1C_2) = P(C_1)P(C_2) = (0.9)(0.9) = 0.81$$

P(current flows in parallel system) = P(at least one of the relays is closed)

$$= P(C_1 \cup C_2) = P(C_1) + P(C_2) - P(C_1C_2)$$

$$= 0.9 + 0.9 - (0.9)(0.9) = 0.99$$

2.39 Let F denote the event that a worker fails to learn the skill correctly.

$$P(A|F) = \frac{P(A)P(F|A)}{P(F|A)P(A) + P(F|B)P(B)}$$

$$= \frac{(0.70)(0.20)}{(0.20)(0.70) + (0.10)(0.30)} = 0.8235$$

2.41 **a.** $P(P|M) = \frac{24}{24+16} = \frac{3}{5}$

b. $P(M|P) = \frac{24}{24+36} = \frac{2}{5}$

c. $P(PM) = \frac{24}{100} = \left(\frac{24+36}{100}\right)\left(\frac{24+16}{100}\right) = P(P)P(M)$. Therefore, events P and M are independent.

d. $P(PF) = \frac{36}{100} = \left(\frac{60}{100}\right)\left(\frac{36+24}{100}\right) = P(P)P(F)$. Therefore, events P and F are independent.

2.43 $P(A)P(B|A)P(C|AB) = P(A) \cdot \frac{P(AB)}{P(A)} \cdot \frac{P(ABC)}{P(AB)} = P(ABC)$

2.45 **a.** P(the defective item is in box i) = $\frac{1}{k}$. Our assumption is that the boxes were filled independently of one another.

b. Let A = the event that the defective item is found in box 1 and B = the event that the defective item was placed in box 1. Then, $P(A|B) = \frac{n}{m}$

c. $P(A^c) = P(A^c|B)P(B) + P(A^c|B^c)P(B^c)$

$$= \left(\frac{m-n}{m}\right)\left(\frac{1}{k}\right) + (1)\left(\frac{k-1}{k}\right) = \frac{m-n}{mk} + \frac{k-1}{k}$$

$$= \frac{m-n+mk-m}{mk} = \frac{mk-n}{mk} = 1 - \frac{n}{mk}$$

d. $P(B|A^c) = \dfrac{P(BA^c)}{P(A^c)} = \dfrac{P(B)P(A^c|B)}{P(A^c)} = \dfrac{\dfrac{m-n}{mk}}{\dfrac{mk-n}{mk}} = \dfrac{m-n}{mk-n}$

e. Let C = the event that the defective item was placed in box 2.

$$P(C|A^c) = \frac{P(A^cC)}{P(A^c)} = \frac{P(C)P(A^c|C)}{P(A^c)}$$

$$= \frac{\left(\frac{1}{k}\right)(1)}{\dfrac{mk-n}{mk}} = \frac{m}{mk-n}$$

f. $\lim\limits_{n\to m} P(A|B) = \lim\limits_{n\to m} \dfrac{n}{m} = 1$

$$\lim_{n\to m} P(A^c) = \lim_{n\to m} 1 - \frac{n}{mk} = 1 - \frac{1}{k}$$

$$\lim_{n\to m} P(B|A^c) = \lim_{n\to m} \frac{m-n}{mk-n} = 0$$

$$\lim_{n\to m} P(C|A^c) = \lim_{n\to m} \frac{m}{mk-n} = \frac{1}{k-1}$$

And

$$\lim_{n\to 0} P(A|B) = \lim_{n\to 0} \frac{n}{m} = 0$$

$$\lim_{n\to 0} P(A^c) = \lim_{n\to 0} 1 - \frac{n}{mk} = 0$$

$$\lim_{n\to 0} P(B|A^c) = \lim_{n\to 0} \frac{m-n}{mk-n} = \frac{1}{k}$$

$$\lim_{n\to 0} P(C|A^c) = \lim_{n\to 0} \frac{m}{mk-n} = \frac{1}{k}$$

2.47 Assume first that $n \le M - W$ and $n \le W$.

a. $P(A_1) = \dfrac{W}{M}$

$$P(A_2) = P(A_2|A_1)P(A_1) + P(A_2|A_1^c)P(A_1^c)$$

$$= \frac{W-1}{M-1}\,\frac{W}{M} + \frac{W}{M-1}\,\frac{M-W}{M} = \frac{W}{(M-1)M}(W-1+M-W)$$

$$= \frac{W}{(M-1)M}(M-1) = \frac{W}{M}$$

$$P(A_3) = P(A_3|A_2A_1)P(A_2A_1) + P(A_3|A_2A_1^C)P(A_2A_1^C)$$

$$+ P(A_3|A_2^CA_1)P(A_2^CA_1) + P(A_3|A_2^CA_1^C)P(A_2^CA_1^C)$$

$$= \frac{W-2}{M-2}\,\frac{W-1}{M-1}\,\frac{W}{M} + \frac{W-1}{M-2}\,\frac{W}{M-1}\,\frac{M-W}{M}$$

$$+ \frac{W-1}{M-2}\,\frac{M-W}{M-1}\,\frac{W}{M} + \frac{W}{M-2}\,\frac{M-W-1}{M-1}\,\frac{M-W}{M}$$

$$= \frac{W}{(M-2)(M-1)M}((W-2)(W-1) + 2(W-1)(M-W) + (M-W-1)(M-W))$$

$$= \frac{W}{(M-2)(M-1)M}(W^2 - 3W + 2 + 2MW - 2M + 2W - 2W^2 + M^2 - 2MW - M - W + W^2)$$

$$= \frac{W}{(M-2)(M-1)M}(M^2 - 3M + 2)$$

$$= \frac{W}{(M-2)(M-1)M}(M-1)(M-2) = \frac{W}{M}$$

Therefore, $P(A_j) = \frac{W}{M} = p$

b. $P(A_j|B_k) = \frac{P(A_jB_k)}{P(B_k)}$

$$= \frac{P(B_k|A_j)P(A_j)}{P(B_k)} = \frac{\binom{n-1}{k-1}p^{k-1}(1-p)^{n-k}\cdot p}{\binom{n}{k}p^k(1-p)^{n-k}} = \frac{k}{n}$$

(Note similarity with a.)

c. No; replace binomial with hypergeometric.

2.49 **a.**

	Yes	No	Total
White defendant	19	141	160
Black defendant	17	149	166
Total	36	290	426

Odds Ratio:

$$\left(\frac{19}{141}\right) \Big/ \left(\frac{17}{149}\right) = 1.18$$

b.

White victim	Yes	No	Total
White defendant	19	132	151
Black defendant	11	52	63
Total	30	184	214

Odds Ratio:

$$\left(\frac{19}{132}\right) \Big/ \left(\frac{11}{52}\right) = 0.68$$

Black victim	Yes	No	Total
White defendant	0	9	9
Black defendant	6	97	103
Total	6	106	112

Odds Ratio:

$$\left(\frac{0}{9}\right) \Big/ \left(\frac{6}{97}\right) = 0$$

c. Part (b) does not imply part (a). The apparent paradox is that we are trying to use a percentage based on one category on all categories.

2.51 Let the outcome of the four tosses be denoted by an ordered quartet of letters, in which the ith letter of the quartet is H if the ith toss resulted in heads, or T if the ith toss resulted in tails, $i = 1, 2, 3, 4$.

a. The sixteen outcomes for the experiment are

$HHHH$ $\quad$ $THHH$

$HHHT$ $\quad$ $THHT$

$HHTH$ $\quad$ $THTH$

$HHTT$ $\quad$ $THTT$

$HTHH$ $\quad$ $TTHH$

$HTHT$ $\quad$ $TTHT$

$HTTH$ $\quad$ $TTTH$

$HTTT$ $\quad$ $TTTT$

b. $A = \{HHHT,\ HHTH,\ HTHH,\ THHH\}$

c. Assuming that all outcomes are equally likely, we have that the probability of each outcome is 1/16 and

$$P(A) = 4 \cdot \frac{1}{16} = \frac{1}{4}.$$

2.53 **a.** Let C_1 = event of a car traveling at less than 55 mph on a rural interstate.

Year	$P(C_1)$
1973	$0.05 + 0.02 = 0.07$
1974	$0.24 + 0.07 + 0.01 = 0.32$
1975	$0.23 + 0.05 + 0.01 = 0.29$

b. Let T = event of a truck traveling at less than 55 mph on a rural interstate.

Year	$P(T)$
1973	$0.15 + 0.05 + 0.02 + 0.01 = 0.23$
1974	$0.29 + 0.11 + 0.02 + 0.01 = 0.43$
1975	$0.29 + 0.08 + 0.02 = 0.39$

c. Let C_2 = event of a car traveling at less than 55 mph on a rural secondary road.

Year	$P(C_2)$
1973	$0.19 + 0.19 + 0.11 + 0.06 + 0.04 = 0.59$
1974	$0.23 + 0.25 + 0.14 + 0.09 + 0.04 = 0.75$
1975	$0.26 + 0.21 + 0.10 + 0.06 + 0.02 = 0.65$

2.55 $\binom{3}{0} + \binom{3}{1} + \binom{3}{2} + \binom{3}{3} = 1 + 3 + 3 + 1 = 8 = 2^3$

Recall that

$$(a+b)^n = \sum_{i=0}^{n} \binom{n}{i} a^i b^{n-i}.$$

Therefore,

$$(1+1)^n = \sum_{i=0}^{n} \binom{n}{i} 1^i 1^{n-i}; \text{ i.e., } 2^n = \sum_{i=0}^{n} \binom{n}{i}.$$

2.57 $\binom{10}{3} = \frac{10!}{7!3!} = \frac{10 \cdot 9 \cdot 8}{3 \cdot 2} = 120$

2.59 $P_3^{10} = 10 \cdot 9 \cdot 8 = 720$

2.61 $P_8^8 = 8! = 40,320$

2.63 Denote success on try i as S_i, failure as F_i, $i = 1, 2, 3$.

a. $P(S_1S_2S_3) = P(S_1)P(S_2)P(S_3) = (0.6)^3 = 0.216$

b. $P(\text{at least one success}) = 1 - P(\text{no successes}) = 1 - P(F_1F_2F_3)$

$= 1 - P(F_1)P(F_2)P(F_3) = 1 - (0.4)^3 = 0.936$

c. $P(\text{at least two successes}) = P(\text{two successes}) + P(\text{three successes})$

$$= P(S_1S_2F_3 \cup S_1F_2S_3 \cup F_1S_2S_3) + P(\text{three successes})$$

$$= P(S_1S_2F_3) + P(S_1F_2S_3)$$

$$+ P(F_1S_2S_3) + P(\text{three successes})$$

$$= 3(0.6)(0.6)(0.4) + (0.6)^3 = 0.648$$

2.65 Using the notation and results of Exercise 2.64, we have

$P(\text{match occurs for first time on fourth toss})$

$$= P(\overline{M}_1\overline{M}_2\overline{M}_3M_4) = P(\overline{M}_1)P(\overline{M}_2)P(\overline{M}_3)P(M_4)$$

$$= [1 - P(M_i)]^3 P(M_i) = \left(1 - \frac{1}{2}\right)^3 \frac{1}{2} = \left(\frac{1}{2}\right)^4 = \frac{1}{16}$$

2.67 Denote a win for gambler Jones on toss i as W_i, and a loss as L_i, $i = 1, 2, \ldots.$

Then $P(W_i) = P(L_i) = 1/2$.

a. $P(\text{break even after 6 tosses}) = P(\text{3 losses and 3 wins in 6 tosses})$

$$= \begin{pmatrix} \text{number of ways of} \\ \text{drawing 3 from 6} \end{pmatrix} P \begin{pmatrix} \text{throw 3 wins and} \\ \text{3 losses in a} \\ \text{given sequence} \end{pmatrix}$$

$$= \binom{6}{3}\left(\frac{1}{2}\right)^6 = 20\left(\frac{1}{64}\right) = \frac{5}{16}$$

b. $P(\text{Jones wins on } 10^{th} \text{ toss})$

$$= P\left\{ [\underbrace{(\text{5 wins and 1 loss})}_{\text{1st 6 tosses}}] \cap \underbrace{(\text{1 win and 1 loss})}_{\text{next 2 tosses}} \right.$$

$$\left. \cap \underbrace{(\text{2 wins})}_{\text{9th and 10th tosses}}] \cup [\underbrace{(\text{4 wins and 2 losses})}_{\text{1st 6 tosses}}] \cap \underbrace{(\text{4 wins})}_{\text{next 4 tosses}}] \right\}$$

$$\left[\binom{6}{5}\left(\frac{1}{2}\right)^6 \cdot \binom{2}{1}\left(\frac{1}{2}\right)^2 \cdot \frac{1}{2} \cdot \frac{1}{2}\right] + \left[\binom{6}{4}\left(\frac{1}{2}\right)^6 \cdot \left(\frac{1}{2}\right)^4\right]$$

$$= \left(\frac{1}{2}\right)^{10}(12 + 15) = 27\left(\frac{1}{2}\right)^{10}$$

2.69 $P(A) > 0$, and $P(B) > 0$, but $P(AB) = 0 \neq P(A)P(B)$ since A and B are mutually exclusive. Therefore, A and B are not independent.

2.71 Denote the event of a defective fan by D. We have $P(A) = 0.9$, $P(B) = 0.1$, $P(AB) = 0$, $P(D|A) = 0.05$, $P(D|B) = 0.03$. Then

$$P(B|D) = \frac{P(D|B)P(B)}{P(D|B)P(B) + P(D|A)P(A)}$$

$$= \frac{(0.03)(0.1)}{(0.03)(0.01) + (0.05)(0.9)} = \frac{1}{16} = 0.0625$$

2.73 $P(A\bar{B}) = P(A) - P(AB)$

$= P(A) - P(A)P(B)$ (because of independence)

$= P(A)[1 - P(B)] = P(A)P(\bar{B})$

2.75 In order to form a triangle from the line seqments ax, bx, and ac we must have the following:

$$ax + ac > bx \quad \text{for} \quad a < x < c$$

$$bx + ac > ax \quad \text{for} \quad c < x < b$$

i.e., for $a < x < c$:

$$(x - a) + \frac{b-a}{2} > (b - x)$$

i.e., $$x > \frac{1}{2}\left(b + a - \frac{b-a}{2}\right)$$

i.e., $$x > c - \frac{b-a}{4}$$

and, for $c < x < b$:

$$(b - x) + \frac{b-a}{2} > (x - a)$$

i.e., $$x < \frac{1}{2}\left(b + a + \frac{b-a}{2}\right)$$

i.e., $$x < c + \frac{b-a}{4}.$$

Therefore, we must have

$$\left(c - \frac{b-a}{4}\right) < x < \left(c + \frac{b-a}{4}\right)$$

i.e., x must fall within a segment half the length ab. Therefore, the desired probability is 1/2.

2.77 $N - 1$ physical dividers are required in order to divide n balls into N groups (cells or boxes), $n \geq N$. Therefore, the number of ways of arranging n balls into N cells is equal to the number of ways of choosing n units out of $(N + n - 1)$ units (the n chosen units are declared to be "balls" and the remaining $N - 1$ units are declared to be "dividers"); i.e., there are

$$\binom{N+n-1}{n} = \binom{N+n-1}{N-1}$$

ways of allocating n balls to N cells.

If no cell is empty, then we have $n - N$ balls whose placement is unspecified, and following the above reasoning, there are

$$\binom{(n-N)+N-1}{n-N} = \binom{n-1}{N-1}$$

ways of allocating these $n - N$ balls to N cells.

Therefore,

$$P\begin{pmatrix}\text{no box will} \\ \text{be empty}\end{pmatrix} = \frac{\begin{pmatrix}\text{number of ways of} \\ \text{allocating } n-N \text{ balls} \\ \text{to } N \text{ cells}\end{pmatrix}}{\begin{pmatrix}\text{number of ways of} \\ \text{allocating } n \text{ balls} \\ \text{to } N \text{ cells}\end{pmatrix}} = \frac{\binom{n-1}{N-1}}{\binom{N+n-1}{N-1}}$$

2.79 Denote the event of person having hepatitis as H, and a positive test result as T. We are given

$P(T \mid H) = 0.9$, $P(T \mid \overline{H}) = 0.01$, and $P(H) = \dfrac{1}{10,000}$.

a. $$P(H \mid T) = \frac{P(T \mid H)P(H)}{P(T \mid H)P(H) + P(T \mid \overline{H})P(\overline{H})}$$

$$= \frac{(0.9)\,(0.0001)}{(0.9)\,(0.0001) + (0.01)\,(0.9999)} = 0.00892$$

b. Here, $P(H) = \dfrac{1}{2}$.

$$P(H \mid T) = \frac{(0.9)\,(0.5)}{(0.9)\,(0.5) + (0.01)\,(0.5)} = 0.9890$$

CHAPTER 3
DISCRETE PROBABILITY DISTRIBUTIONS

3.1 $P(X=0) = P(3 \text{ males chosen})$

$$= \frac{\begin{pmatrix}\text{number of ways of}\\ \text{choosing 3 out of 4}\end{pmatrix}\begin{pmatrix}\text{number of ways of}\\ \text{choosing 0 out of 6}\end{pmatrix}}{\begin{pmatrix}\text{total number of}\\ \text{ways of choosing 3 out of 10}\end{pmatrix}}$$

$$= \frac{\binom{4}{3}\binom{6}{0}}{\binom{10}{3}} = \frac{1}{30}$$

$$P(X=1) = P(2 \text{ males and 1 female chosen}) = \frac{\binom{4}{2}\binom{6}{1}}{\binom{10}{3}} = \frac{3}{10}$$

$$P(X=2) = \frac{\binom{4}{1}\binom{6}{2}}{\binom{10}{3}} = \frac{1}{2}$$

$$P(X=3) = \frac{\binom{4}{0}\binom{6}{3}}{\binom{10}{3}} = \frac{1}{6}$$

3.3 $P(X=0) = \binom{3}{0}(0.363)^0(0.637)^3 = 0.2585$

$P(X=1) = \binom{3}{1}(0.363)^1(0.637)^2 = 0.4419$

$P(X=2) = \binom{3}{2}(0.363)^2(0.637)^1 = 0.2518$

$P(X=3) = \binom{3}{3}(0.363)^3(0.637)^0 = 0.0478$

This answer assumes independence of up-at-bats. This assumption may not be reasonable, since pitchers may change between up-at-bats. Boggs might get tired as the game progresses, etc. It appears that it is not unusual for a good hitter to go 0 for 3 in one game, since the probability of this for Boggs is more than 1/4.

3.5 The probability that a given residential fire is in a family home is 0.73.

a.

x	$p(x)$		
0	$\binom{4}{0}(0.27)^4$	=	0.0053
1	$\binom{4}{1}(0.27)^3(0.73)$	=	0.0575
2	$\binom{4}{2}(0.27)^2(0.73)^2$	=	0.2331
3	$\binom{4}{3}(0.27)\ (0.73)^3$	=	0.4201
4	$\binom{4}{4}(0.73)^4$	=	0.2840

b. $P(X \geq 1) = 1 - P(X = 0) = 1 - 0.0053 = 0.9947$

3.7

$$p(x) = \binom{3}{x}\left(\frac{1}{3}\right)^{x}\left(\frac{2}{3}\right)^{3-x} ; x = 0, 1, 2, 3$$

$$p(y) = \binom{3}{y}\left(\frac{1}{15}\right)^{y}\left(\frac{14}{15}\right)^{3-y} ; y = 0, 1, 2, 3$$

x	$p(x)$	y	$p(y)$
0	8/27	0	2744/3375
1	12/27	1	588/3375
2	6/27	2	42/3375
3	1/27	3	1/3375

$$\begin{aligned}
P(X+Y=0) &= P(X=0)P(Y=0) \\
&= \frac{8}{27}\cdot\frac{2744}{3375} = 0.24090 \\
P(X+Y=1) &= P(X=0)P(Y=1) + P(X=1)P(Y=0) \\
&= \frac{8}{27}\cdot\frac{588}{3375} + \frac{12}{27}\cdot\frac{2744}{3375} = 0.41297 \\
P(X+Y=2) &= P(X=1)P(Y=1) + P(X=2)P(Y=0) + P(Y=2)P(X=0) \\
&= \frac{12}{27}\cdot\frac{588}{3375} + \frac{6}{27}\cdot\frac{2744}{3375} + \frac{8}{27}\cdot\frac{42}{3375} = 0.26179 \\
P(X+Y=3) &= P(X=0)P(Y=3) + P(X=1)P(Y=2) \\
&\quad + P(X=2)P(Y=1) + P(X=3)P(Y=0) \\
&= \frac{8}{27}\cdot\frac{1}{3375} + \frac{12}{27}\cdot\frac{42}{3375} + \frac{6}{27}\cdot\frac{588}{3375} + \frac{1}{27}\cdot\frac{2744}{3375} = 0.07445 \\
P(X+Y=4) &= P(X=1)P(Y=3) + P(X=2)P(Y=2) + P(X=3)P(Y=1) \\
&= \frac{12}{27}\cdot\frac{1}{3375} + \frac{6}{27}\cdot\frac{2744}{3375} + \frac{1}{27}\cdot\frac{588}{3375} = 0.00935 \\
P(X+Y=5) &= P(X=2)P(Y=3) + P(X=3)P(Y=2) \\
&= \frac{6}{27}\cdot\frac{1}{3375} + \frac{1}{27}\cdot\frac{42}{3375} = 0.00053 \\
P(X+Y=6) &= P(X=3)P(Y=3) \\
&= \frac{1}{27}\cdot\frac{1}{3375} = 0.00001
\end{aligned}$$

3.9 **a.** $$p(x) = \frac{\begin{pmatrix}\text{number of ways of}\\ \text{choosing } x \text{ from } 2\end{pmatrix}\begin{pmatrix}\text{number of ways of}\\ \text{choosing } 2-x \text{ from } 2\end{pmatrix}}{\begin{pmatrix}\text{total number of ways of}\\ \text{choosing a sample of 2 from 4}\end{pmatrix}}$$

$$\frac{\binom{2}{x}\binom{2}{2-x}}{\binom{4}{2}} \quad x = 0, 1, 2;$$

i.e,

x	$p(x)$
0	1/6
1	2/3
2	1/6

b. $$p(x) = \frac{\binom{1}{x}\binom{3}{2-x}}{\binom{4}{2}} \quad x = 0, 1;$$

i.e.,

x	$p(x)$
0	1/2
1	1/2

c. $P(X = 0) = 1$

3.11 Let X = number on the ticket drawn, and G_i = net gain for box i, i = I, II. Then $G_i = X - 1$.

a. $$E(G_I) = \sum_{x=0}^{2} (x-1)\,p_I(x) = (-1)\left(\frac{1}{3}\right) + 0\left(\frac{1}{3}\right) + 1\left(\frac{1}{3}\right) = 0$$

$$E(G_I^2) = \sum_{x=0}^{2} (x-1)^2 p_I(x) = 1\left(\frac{1}{3}\right) + 0\left(\frac{1}{3}\right) + 1\left(\frac{1}{3}\right) = \frac{2}{3}$$

$$V(G_I) = E(G_I^2) - [E(G_I)]^2 = \frac{2}{3} - 0 = \frac{2}{3}$$

b. $E(G_{II}) = \sum_{x=0}^{2} (x-1)p_{II}(x) = (-1)\left(\frac{3}{5}\right) + 0\left(\frac{1}{5}\right) + 3\left(\frac{1}{5}\right) = 0$

$$E(G_{II}^2) = \sum_{x=0}^{2} (x-1)^2 p_{II}(x) = 1\left(\frac{3}{5}\right) + 0\left(\frac{1}{5}\right) + 9\left(\frac{1}{5}\right) = \frac{12}{5}$$

$$V(G_{II}) = E(G_{II}^2) - [E(G_{II})]^2 = \frac{12}{5} - 0 = \frac{12}{5}$$

c. Box II, since for Box I the highest possible net gain is \$1 with a probability of 1/3, but for Box II the highest possible gain is \$3 with probability of 1/5. Note that $V(G_I) < V(G_{II})$; i.e., Box I net gain varies less from $E(G_i) = 0$ than Box II net gain.

3.13 Let X = age of death of a person infected with the AIDS virus from 1982–1989. Then we can estimate the mean and standard deviation of X from Figure 3.5 by letting the possible values of X be the mean ages for the different age groups in the pie chart. Hence,

$E(X) = 6P(X = 6) + 21P(X = 21) + 34.5P(X = 34.5)$

$+ 44.5P(X = 44.5) + 54.5P(X = 54.5) + 60P(X = 60)$

$= 6(0.014) + 21(0.194) + 34.5(0.45) + 44.5(0.221) + 54.5(0.082) + 60(0.038)$

$= 36.3$ (Answers may vary.)

$\text{Var}(X) = E(X^2) - (E(X))^2$

$E(X^2) = 36(0.014) + 441(0.194) + 1190.25(0.45) + 1980.25(0.221)$

$+ 2970(0.082) + 3600(0.038) = 1439.65$

$\text{Var}(X) = 1439.65 - (36.3)^2 = 1439.65 - 1317.69 = 121.96$ (Answers may vary.)

$\text{Std}(X) = (\text{Var}(X))^{(1/2)} = (121.96)^{(1/2)} = 11.04$ (Answers may vary.)

3.15 $E(\text{number of sales}) = 0 \cdot p(0) + 1 \cdot p(1) + 2 \cdot p(2)$

$= 0(0.7) + 1(0.2) + 2(0.1) = 0.4$

$V(\text{number of sales}) = 0^2 \cdot p(0) + 1^2 \cdot p(1) + 2^2 \cdot p(2) - [E(\text{number of sales})]^2$

$= 0(0.7) + 1(0.2) + 4(0.1) - (0.4)^2 = 0.44$

Standard deviation of sales $= \sqrt{V(\text{sales})} = \sqrt{0.44} = 0.6633$

3.17 Let X = number of contracts awarded to firm I, and let A_i = event that firm I gets contract i, $i = 1, 2$.

x	$p(x)$	
0	$P(\overline{A_1}\overline{A_2})$	$= \frac{2}{3} \cdot \frac{2}{3} = \frac{4}{9}$
1	$P(A_1\overline{A_2}) + P(\overline{A_1}A_2)$	$= 2 \cdot \frac{1}{3} \cdot \frac{2}{3} = \frac{4}{9}$
2	$P(A_1A_2)$	$= \frac{1}{3} \cdot \frac{1}{3} = \frac{1}{9}$

$$E(X) = 0 \cdot p(0) + 1 \cdot p(1) + 2 \cdot p(2) = 0 \cdot \frac{4}{9} + 1 \cdot \frac{4}{9} + 2 \cdot \frac{1}{9} = \frac{2}{3}$$

Let $Y = 90{,}000X$. Then, by Theorem 3.2, $E(Y) = 90{,}000E(X) = 90{,}000(2/3) = \$60{,}000$.

Let X_i = the number of contracts assigned to firm i, $i = 1, 2$. Note that the distribution of X_1 is the same as that of X_2. Let Z = profit for both firms = $90{,}000(X_1 + X_2)$. Then, by Theorem 3.2, we have

$$E(Z) = 90,000E(X_1 + X_2) = 90,000[E(X_1) + E(X_2)]$$

$$= 90,000\left(\frac{2}{3} + \frac{2}{3}\right) = \$120{,}000$$

3.19 **a.** Let X = weekly number of breakdowns. From Tchebysheff's theorem, we have

$$P(\mu - k\sigma < X < \mu + k\sigma) \geq 1 - \frac{1}{k^2}$$

For $1 - \frac{1}{k_2} = 0.9$, we have $k = \sqrt{10}$, and thus the desired interval is $(\mu - k\sigma, \mu + 5\sigma) = [4 - \sqrt{10}(0.8), 4 + \sqrt{10}(0.8)] = (1.4702, 6.5298)$.

b. Eight breakdowns is $\frac{8 - \mu}{\sigma} = \frac{8 - 4}{0.8} = 5$ standard deviations from the mean. The interval $(\mu - 5\sigma, \mu + 5\sigma)$ or $(0, 8)$ must contain at least $1 - \frac{1}{5^2} = 0.96$ of the probability. Thus, at most 4% of the probability mass can exceed 8 breakdowns and the director is safe in his claim.

3.21 **a.** Let X = battery performance period. From Tchebysheff's theorem, we have

$$P(\mu - k\sigma < X < \mu + k\sigma) \geq 1 - \frac{1}{k^2}$$

Solving $1 - \frac{1}{k^2} = 0.9$, for $k = \sqrt{10}$; thus the desired interval is

$$(\mu - \sqrt{10}\sigma, \mu + \sqrt{10}\sigma) = [100 - \sqrt{10}(5), 100 + \sqrt{10}(5)] = (84.1886, 115.8114).$$

b. No, since $80 \not\subset (84.1886, 115.8114)$. Also, note that 80 is $\frac{80 - 100}{5} = -4$ standard deviations from the mean. Then $P(X \leq \mu - 4\sigma) \leq P(|X - \mu| \geq 4\sigma) \leq \frac{1}{4^2} = 0.0625$. Therefore, one would expect less than 6.25% of batteries to die out in less than 80 minutes.

3.23 **a.** $P(X=2) = \binom{4}{2}(0.2)^2(0.8)^2 = 0.1536$

b. $P(X \geq 2) = P(X=2) + P(X=3) + P(X=4)$

$$= \sum_{x=2}^{4} \binom{4}{x}(0.2)^x(0.8)^{4-x}$$

$$= \binom{4}{2}(0.2)^2(0.8)^2 + \binom{4}{3}(0.2)^3(0.8) + \binom{4}{4}(0.2)^4$$

$$= 0.1536 + 0.0256 + 0.0016 = 0.1808.$$

c. $P(X \leq 2) = P(X=0) + P(X=1) + P(X=2)$

$$= \sum_{x=0}^{2} \binom{4}{x}(0.2)^x(0.8)^{4-x}$$

$$= \binom{4}{0}(0.8)^4 + \binom{4}{1}(0.2)^1(0.8)^3 + \binom{4}{2}(0.2)^2(0.8)^2$$

$$= 0.4096 + 0.4096 + 0.1536 = 0.9728$$

d. $E(X) = np = 4(0.2) = 0.8$

e. $V(X) = np(1-p) = 4(0.2)(0.8) = 0.64$

3.25 Let X = number of underfilled boxes. Then X has a binomial distribution with parameters $n = 25$, p as given.

a. $P(X \leq 2) = 0.537$

b. $P(X \leq 2) = 0.098$

3.27 **a.** $E(X) = np = 20(0.8) = 16$

b. $V(X) = np(1-p) = 20(0.8)(0.2) = 3.2$

3.29 $P(X \geq 5) = 1 - P(X \leq 4) > 0.9 \Rightarrow P(X \leq 4) < 0.1$

Using the formula

$$P(X \leq 4) = \sum_{x=0}^{4} \binom{n}{x}(0.8)^x(0.2)^{n-x}$$

we find

n	$P(X \le 4)$
6	0.34464
7	0.14803
8	0.05628

Therefore, at least $n = 8$ people must donate blood to have the probability of at least 5 Rh+ donors be greater than 0.9.

3.31 Let X = number of firms out of a sample of five that say "quality of life" is an important factor. Then, assuming independence among firms, X has a binomial distribution with parameters $n = 5$, $p = 0.55$, and

$$P(X \ge 3) = \sum_{x=3}^{5} \binom{5}{x} (0.55)^x (0.45)^{5-x}$$

$$= \binom{5}{3}(0.55)^3(0.45)^2 + \binom{5}{4}(0.55)^4(0.45)^1 + \binom{5}{5}(0.55)^5$$

$$= 0.3369 + 0.2059 + 0.0503 = 0.5931$$

3.33 Let X = number of radar sets out of n that detect an intruding aircraft. Then X has a binomial distribution with parameters n, and $p = 0.9$.

a. $P(X \ge 1) = 1 - P(X = 0) = 1 - \binom{2}{0}(0.9)^0(0.1)^2 = 0.99$

b. $P(X \ge 1) = 1 - P(X = 0) = 1 - \binom{4}{0}(0.9)^0(0.1)^4 = 0.9999$

3.35 Let X = number of components out of the four that last longer than 1000 hours. The probability that a given component lasts longer than 1000 hours is 0.8; thus X has a binomial distribution with parameters $n = 4$, $p = 0.8$.

a. $P(X = 2) = \binom{4}{2}(0.8)^2(0.2)^2 = 0.1536$

b. $P(X \ge 2) = 1 - P(X \le 1) = 1 - \binom{4}{0}(0.8)^0(0.2)^4 - \binom{4}{1}(0.8)^1(0.2)^3$

$= 1 - 0.0016 - 0.0256 = 0.9728$

3.37 Y has a binomial distribution with parameters $n = 4$, $p = 0.1$.

$$E(C) = E(3Y^2 + Y + 2) = 3E(Y^2) + E(Y) + 2$$

$$= 3(V(Y) + [E(Y)]^2) + E(Y) + 2 = 3np(1-p) + 3(np)^2 + np + 2$$

$$= 3(4)(0.1)(0.9) + 3[(4)(0.1)]^2 + 4(0.1) + 2 = 3.96$$

3.39 Let X = number of defective motors out of ten in the warehouse. Then X is binomially distributed with $n = 10, p = 0.08$. Let Y = net gain = (selling price for the ten motors) – (twice the selling price of a motor) × $X = 10(100) - 200X = 1{,}000 - 200X$.

$$E(Y) = E(1{,}000 - 200X) = 1{,}000 - 200E(X)$$

$$= 1{,}000 - 200np = 1{,}000 - 200(10)(0.08) = 840$$

3.41 **a.** $P(Y \geq 4) = 1 - P(Y \leq 3) = 1 - P(Y = 2) - P(Y = 3)$

$$= 1 - \binom{2-1}{2-1}(0.4)^2(0.6)^0 - \binom{3-1}{2-1}(0.4)^2(0.6)^1$$

$$= 1 - 0.16 - 2(0.4)^2(0.6) = 0648$$

b. $P(Y = y)$ is non-zero only for $y = r, r+1, r+2, \ldots$ Therefore, for $r = 4$,

$$P(Y \geq 4) = \sum_{y=4}^{\infty} p(y) = 1.$$

3.43 Let X = the trial on which the third nondefective engine is found. Then X has a negative binomial distribution, with $p = 0.9, r = 3$.

a. $P(X = 5) = p(5) = \binom{y-1}{r-1} p^r (1-p)^{y-r} = \binom{4}{2}(0.9)^3(0.1)^2 = 0.04374$

b. $P(X \leq 5) = P(X = 3) + P(X = 4) + P(X = 5) = p(3) + p(4) + p(5)$

$$= \binom{3-1}{3-1}(0.9)^3(0.1)^0 + \binom{4-1}{3-1}(0.9)^3(0.1)^1$$

$$+ \binom{5-1}{3-1}(0.9)^3(0.1)^2$$

$$= 0.729 + 0.2187 + 0.04374 = 0.99144$$

3.45 **a.** Let Y be defined as in Exercise 3.42. Then Y has a geometric distribution with $p = 0.9$, and

$$E(Y) = \frac{1}{p} = \frac{10}{9}$$

$$V(Y) = \frac{1-p}{p^2} = \frac{0.1}{(0.9)^2} = 0.1235.$$

b. Let X be defined as in Exercise 3.43. Then X has a negative binomial distribution with parameters $p = 0.9, r = 3$, and

$$E(X) = \frac{r}{p} = \frac{30}{9} = 3.33$$

$$V(X) = \frac{r(1-p)}{p^2} = \frac{0.3}{(0.9)^2} = 0.3704.$$

3.47 Let total cost = C. Then $C = 20X$, and

$$E(C) = 20E(X) = 20\left(\frac{r}{p}\right) = 20\left(\frac{3}{0.4}\right) = 150$$

$$V(C) = V(20X) = 20^2 V(X) = 400\frac{r(1-p)}{p^2} = 400\frac{(3)(0.6)}{(0.4)^2} = 4,500$$

Standard deviation of $C = \sqrt{V(C)} = \sqrt{4,500} = 67.082$

Using Tchebysheff's theorem, we have $P(C > 350) = P(C - 150 > 350 - 150) \leq P(|C - 150| \geq 200)$

$= P\left(|C - 150| \geq \left(\frac{200}{67.082}\right)67.082\right) \leq \frac{1}{\left(\frac{200}{67.082}\right)^2} = 0.1125$. Therefore, it is unlikely that the cost will exceed \$350.

Note, also that $P(C > 350)$ may be computed exactly as

$$P(C > 350) = P(20X > 350) = P(X > 17.5)$$

$$= 1 - P(X \leq 17) = 1 - \sum_{x=3}^{17}\binom{x-1}{3-1}(0.4)^3(1-0.4)^{x-3}.$$

3.49 **a.** Let X = number of the well on which oil was first struck. Then X has a geometric distribution with parameter $p = 0.2$, so

$$P(X = 3) = (1-p)^{(3-1)}p^1 = (0.8)^2(0.2)^1 = 0.128.$$

b. Let Y = number of the well on which the third oil strike occurs. Then Y has a negative binomial distribution with parameters $p = 0.2$, $r = 3$, so

$$P(X = 5) = \binom{5-1}{3-1}(0.2)^3(0.8)^2 = 0.03072.$$

The solutions to parts (a) and (b) require the assumption of independence among wells.

3.51 Let Y = number of tires that must be selected in order to get four good ones. Then Y has a negative binomial distribution with parameters $p = 0.9$, $r = 4$.

a. $P(Y = 6) = \binom{6-1}{4-1}(0.9)^4(0.1)^2 = 0.06561$

b. $E(Y) = \frac{r}{p} = \frac{4}{0.9} = 4.4444$

$$V(Y) = \frac{r(1-p)}{p^2} = \frac{4(0.1)}{(0.9)^2} = 0.4938$$

3.53 **a.** Let Y = number of customers it takes to sell the three white appliances. Then Y has a negative binomial distribution with parameters $p = \frac{1}{2}$, $r = 3$, and

$$P(Y = 5) = \binom{5-1}{3-1}\left(\frac{1}{2}\right)^3\left(\frac{1}{2}\right)^{5-3} = 6\left(\frac{1}{8}\right)\left(\frac{1}{4}\right) = \frac{3}{16}.$$

b. Let X = number of customers it takes to sell the brown appliances. Then X has the same distribution as Y, a negative binomial distribution with parameters $p = \frac{1}{2}$, $r = 3$, and $P(X = 5) =$ $P(Y = 5) = \frac{3}{16}$.

c. $P(Y = 3) = \binom{3-1}{3-1}\left(\frac{1}{2}\right)^3\left(\frac{1}{2}\right)^{3-3} = \left(\frac{1}{2}\right)^3 = \frac{1}{8}$

d. P(all the whites ordered before all browns) =

$$P(Y \le 5) = p(3) + p(4) + p(5) = \frac{3}{16} + p(4) + \frac{1}{8}$$

$$= \frac{3}{16} + \binom{4-1}{3-1}\left(\frac{1}{2}\right)^3\left(\frac{1}{2}\right) + \frac{1}{8}$$

$$= \frac{3}{16} + \frac{3}{16} + \frac{1}{8} = \frac{1}{2}$$

3.55 Let Y = number of calls arriving in a given one-minute period. Then Y has a Poisson distribution with parameter $\lambda = 4$.

a. $P(Y = 0) = p(0) = \frac{4^0}{0!}e^{-4} = e^{-4} = 0.0183$

b. $P(Y \ge 2) = 1 - P(Y \le 1) = 1 - F(1) = 1 - 0.092 = 0.908$

c. Let X = number of calls arriving in a given two-minute period. Then X has a Poisson distribution with parameter $\lambda = 2(4) = 8$, and $P(X \ge 2) = 1 - F(1) = 1 - 0.003 = 0.997$.

3.57 Let Y = number of fatalities per 10^9 vehicle miles with NMSL in effect. Then Y has a Poisson distribution with parameter $\lambda = 16$.

a. $P(Y \le 15) = F(15) = 0.467$

b. $P(Y \ge 20) = 1 - P(Y \le 19) = 1 - F(19) = 1 - 0.812 = 0.188$

3.59 **a.** Let Y = number of teleport inquiries in one millisecond. Then Y has a Poisson distribution with parameter $\lambda = 0.2$ and

$$P(Y=0) = \frac{(0.2)^0}{0!} e^{-0.2} = e^{-0.2} = 0.8187.$$

b. Let X = number of teleport inquiries in three milliseconds. Then X has a Poisson distribution with parameter $\lambda = 3(0.2) = 0.6$ and

$$P(X=0) = \frac{(0.6)^0}{0!} e^{-0.6} = e^{-0.6} = 0.5488.$$

3.61 Let Y = number of customer arrivals in a given hour. Then Y has a Poisson distribution with $\lambda = 8$.

a. $P(Y=8) = P(Y \le 8) - P(Y \le 7) = 0.593 - 0.453 = 0.140$

b. $P(Y \le 3) = 0.042$

c. $P(Y \ge 2) = 1 - P(Y \le 1) = 1 - 0.003 = 0.997$

3.63 **a.** Let X = number of customers that arrive in a given two-hour period of time. Then X has a Poisson distribution with $\lambda = 2(8) = 16$ and

$$P(X=2) = \frac{16^2}{2!} e^{-16} = 128 e^{-16} = 1.44 \times 10^{-5}.$$

b. The two one-hour time periods are nonoverlapping, and therefore X = total number of customers that arrive in the given two-hour time period has a Poisson distribution with $\lambda = 2(8) = 16$, and, as for part (a), $P(X=2) = 144 \times 10^{-5}$.

Consistent with this answer, note the following. Let Y_1 = number of customers that arrive between 1:00 p.m. and 2:00 p.m. and Y_2 = number of customers that arrive between 3:00 p.m. and 4:00 p.m. Then Y_1 and Y_2 are each distributed as Poisson with $\lambda = 8$ and

$$\begin{aligned} P(Y_1 + Y_2 = 2) &= P(Y_1 = 0, Y_2 = 2) + P(Y_1 = 1, Y_2 = 1) \\ &\quad + P(Y_1 = 2, Y_2 = 0) \\ &= 2 \cdot p(0)p(2) + [p(1)]^2 \\ &= 2\left(\frac{8^0}{0!} e^{-8}\right)\left(\frac{8^2}{2!} e^{-8}\right) + \left(\frac{8^1}{1!} e^{-8}\right)^2 \\ &= 7.2 \times 10^{-6} + 7.2 \times 10^{-6} = 1.44 \times 10^{-5}. \end{aligned}$$

3.65 Let X = number of imperfections in an eight-square-yard sample. Then X has a Poisson distribution with $\lambda = 8(4) = 32$. Let $C = 10X$ = cost of repair. Then

$$E(C) = 10E(X) = 10\lambda = 10(32) = 320$$

$$V(C) = (10)^2 V(X) = 100\lambda = 100(32) = 3{,}200$$

The standard deviation of $C = \sqrt{V(C)} = \sqrt{3,200} = 40\sqrt{2} = 56.5685$

3.67 $$E[Y(Y-1)] = \sum_{y=0}^{\infty} y(y-1)\frac{\lambda^y e^{-\lambda}}{y!} = \lambda^2 e^{-\lambda}\sum_{y=2}^{\infty}\frac{\lambda^{(y-2)}}{(y-2)!}$$

$$= \lambda^2 e^{-\lambda}\sum_{x=0}^{\infty}\frac{\lambda^x}{x!} = \lambda^2 e^{-\lambda} e^{\lambda} = \lambda^2$$

$$V(Y) = E(Y^2) - [E(Y)]^2 = E(Y^2) - E(Y) + E(Y) - [E(Y)]^2$$

$$= E[Y(Y-1)] + E(Y) - [E(Y)]^2$$

$$= \lambda^2 + \lambda - (\lambda)^2 = \lambda$$

3.69 **a.** Let Y = number of cars arriving in the first hour.

$$P(Y \geq 12) = 1 - P(Y \leq 11) = 1 - 0.999 = 0.001$$

b. Let X = the number of cars arriving in a given eight hours. Then X has a Poisson distribution with $\lambda = 8(4) = 32$ and

$$p(X \leq 11) = \sum_{x=0}^{11}\frac{32^x e^{-32}}{x!} = e^{-32}\sum_{x=0}^{11}\frac{32^x}{x!}$$

$$= e^{-32}\left(1 + 32 + \frac{32^2}{2} + \frac{32^3}{6} + \ldots + \frac{32^{11}}{11!}\right)$$

$$= e^{-32}(1.345732 \times 10\ ) = 0.000017.$$

3.71 Let Y = number of nondefectives in sample of 5. Then Y has a hypergeometric distribution with parameters $k = 6$, $n = 5$, $N = 10$, and

$$P(Y = 5) = p(5) = \frac{\binom{6}{5}\binom{4}{0}}{\binom{10}{5}} = \frac{1}{42}.$$

3.73 Let Y = number of local firms selected. Then Y has a hypergeometric distribution with parameters $k = 4$, $n = 3$, $N = 6$.

a. $P(\text{at least one not local}) = P(\text{not all local}) = 1 - P(Y = 3) = 1 - p(3)$

$$= 1 - \frac{\binom{4}{3}\binom{2}{0}}{\binom{6}{3}} = 1 - \frac{4}{20} = \frac{4}{5}$$

b. $$P(Y = 3) = \frac{\binom{4}{3}\binom{2}{0}}{\binom{6}{3}} = \frac{1}{5}$$

It is unlikely that all three minority members were selected at random, since the probability of that happening is only 1/30. Therefore, little confidence can be placed in the foreman's claim.

3.75 Y has a hypergeometric distribution with parameters $N = 10$, $n = 3$, and k as given.

a. $k = 2$:

y	$p(y)$
0	$\frac{\binom{2}{0}\binom{8}{3}}{\binom{10}{3}} = \frac{7}{15}$
1	$\frac{\binom{2}{1}\binom{8}{2}}{\binom{10}{3}} = \frac{7}{15}$
2	$\frac{\binom{2}{2}\binom{8}{1}}{\binom{10}{3}} = \frac{1}{15}$

b. $k = 4$:

y	$p(y)$
0	$\dfrac{\binom{4}{0}\binom{6}{3}}{\binom{10}{3}} = \dfrac{1}{6}$
1	$\dfrac{\binom{4}{1}\binom{6}{2}}{\binom{10}{3}} = \dfrac{1}{2}$
2	$\dfrac{\binom{4}{2}\binom{6}{1}}{\binom{10}{3}} = \dfrac{3}{10}$
3	$\dfrac{\binom{4}{3}\binom{6}{0}}{\binom{10}{3}} = \dfrac{1}{30}$

3.77 Let Y = number of misfiring plugs among the four removed. Then Y has a hypergeometric distribution with $N = 8$, $n = 4$, $k = 2$, and

$$P(Y = 2) = \frac{\binom{2}{2}\binom{6}{2}}{\binom{8}{4}} = \frac{3}{14}.$$

3.79 Let Y = number of accounts past due that the auditor sees. Then Y has a hypergeometric distribution with $N = 8$, $n = 3$, k as given, and

$$P(Y \geq 1) = 1 - P(Y = 0) = 1 - \frac{\binom{k}{0}\binom{8-k}{3}}{\binom{8}{3}} = 1 - \frac{\binom{8-k}{3}}{56}.$$

a. $k = 2$; $P(Y \geq 1) = 1 - \dfrac{\binom{8-2}{3}}{56} = 1 - 20/56 = 9/14$

b. $k = 4$; $P(Y \geq 1) = 1 - \dfrac{\binom{8-4}{3}}{56} = 1 - 4/56 = 13/14$

c. $k = 7$; $P(Y \geq 1) = 1 - \dfrac{\binom{8-7}{3}}{56} = 1 - 0 = 1$, since the auditor must choose at least two past-due accounts.

3.81 Note that Y has a hypergeometric distribution with parameters $N = 20$, $n = 5$, and k as given

a. $k = 0$: $P(Y \leq 1) = 1$

b. $k = 1$: $P(Y \leq 1) = 1$

c. $k = 2$:

$$P(Y \leq 1) = p(0) + p(1) = \frac{\binom{2}{0}\binom{20-2}{5-0}}{\binom{20}{5}} + \frac{\binom{2}{1}\binom{20-2}{5-1}}{\binom{20}{5}}$$

$$= \frac{21}{38} + \frac{15}{38} = \frac{18}{19}$$

d. $k = 3$:

$$P(Y \leq 1) = p(0) + p(1) = \frac{\binom{3}{0}\binom{17}{5}}{\binom{20}{5}} + \frac{\binom{3}{1}\binom{17}{4}}{\binom{20}{5}}$$

$$= \frac{91}{228} + \frac{105}{228} = \frac{49}{57}$$

e. $k = 4$:

$$P(Y \geq 1) = p(0) + p(1) = \frac{\binom{4}{0}\binom{16}{5}}{\binom{20}{5}} + \frac{\binom{4}{1}\binom{16}{4}}{\binom{20}{5}}$$

$$= \frac{1092}{3876} + \frac{1820}{3876} = \frac{728}{969}$$

3.83 Let Y = number of defectives from line I. Then Y has a hypergeometric distribution with parameters $N = 10$, $n = 5$, $k = 4$ and

$$P(Y=2) \quad = \quad \frac{\binom{4}{2}\binom{6}{3}}{\binom{10}{5}} \quad = \quad \frac{10}{21}.$$

3.85 $M(t) \; = \; E(e^{tY})$

$$= \sum_{y=0}^{n} e^{ty}\binom{n}{y}p^{y}(1-p)^{(n-y)} \; = \; \sum_{y=0}^{n}\binom{n}{y}(pe^{t})^{y}(1-p)^{(n-y)} \; = \; [pe^{t}+(1-p)]$$

since, using the binomial theorem, we have

$$\sum_{x=0}^{n}\binom{n}{x}a^{x}b^{(n-x)} \; = \; (a+b)^{n}.$$

Therefore,

$$E(Y) \; = \; M'(0) \; = \; n[pe^{t}+(1-p)]^{n-1}pe^{t}\big|_{t=0} \; = \; np$$

$$E(Y^{2}) \; = \; M''(0)$$

$$= \; (n(n-1)\,[pe^{t}+(1-p)]^{n-1}(pe^{t})^{2}+n\,[pe^{t}+(1-p)]^{n-1}pe^{t})\big|_{t=0}$$

$$= \; n(n-1)p^{2}+np$$

$$V(Y) \; = \; E(Y^{2}) - [E(Y)]^{2}$$

$$= \; n(n-1)p^{2}+np-(np)^{2} \; = \; np(1-p)$$

3.87 $M_{Y}(t) \; = \; E(e^{tY}) \; = \; E(e^{t(aX+b)}) \; = \; E[e^{tb}e^{(at)X}] \; = \; e^{tb}E[e^{(at)X}] \; = \; e^{tb}M_{X}(at)$

3.89 $P(t) \; = \; E(t^{Y})$

$$= \sum_{y=0}^{n} t^{y}\binom{n}{y}p^{y}(1-p)^{n-y} \; = \; \sum_{y=0}^{n}\binom{n}{y}(pt)^{y}(1-p)^{n-y} \; = \; [pt+(1-p)]$$

$$E(Y) \; = \; P'(1) \; = \; (np[pt+(1-p)]^{n-1}\big|_{t=1} = np)$$

$$E[Y(Y-1)] \; = \; P''(1) \; = \; n(n-1)p^{2}[pt+(1-p)]^{n-2}\big|_{t=1} \; = \; n(n-1)p^{2}$$

$$V(Y) \; = \; E(Y^{2}) - [E(Y)]^{2} \; = \; E(Y^{2}) - E(Y) + E(Y) - [E(Y)]^{2}$$

$$= \; [E(Y^{2}) - E(Y)] + E(Y) - [E(Y)]^{2}$$

$$= \; E[Y(Y-1)] + E(Y) - [E(Y)]^{2}$$

$$= \; n(n-1)p^{2}+np-n^{2}p^{2} \; = \; np(1-p)$$

3.91 The weather of the city can be modelled by a Markov chain. Consider the following transition probability matrix:

$$P = \begin{array}{c} \\ 0 \\ 1 \end{array} \begin{array}{c} \begin{array}{cc} 0 & 1 \end{array} \\ \begin{bmatrix} 0.1 & 0.9 \\ 0.3 & 0.7 \end{bmatrix} \end{array}$$

Let state 0 be a "rainy day" and state 1 be a "sunny day." Now, to find the percentage of sunny days, we need the stationary distribution which is found by solving the matrix equation

$$\pi = \pi P$$

$$\Rightarrow [\pi_1 \pi_2] = [\pi_1 \pi_2] \begin{bmatrix} 0.1 & 0.9 \\ 0.3 & 0.7 \end{bmatrix}$$

$$\Rightarrow \begin{cases} \pi_1 = 0.1\pi_1 + 0.3\pi_2 \\ \pi_2 = 0.9\pi_1 + 0.7\pi_2 \end{cases}$$

$$\Rightarrow \pi_2 = 3\pi_1$$

Since π is a distribution, then $\pi_1 + \pi_2 = 1$, so we have $\pi_1 = 1/4$ and $\pi_2 = 3/4$. Therefore, the city has 75% sunny days in the long run.

3.93 For ease of calculation, let

A_t = # of black balls in urn 1 at time t

B_1 = a black ball is drawn from urn 1

B_2 = a black ball is drawn from urn 2

W_1 = a white ball is drawn from urn 1

W_2 = a white ball is drawn from urn 2

Without loss of generality, start at an initial time t_0 when $A_{t_0} = j \;\; 0 \le j \le n$. Then the transition probabilities can be easily deduced.

$$p_{jj-1} = P[A_{t_0+1} = j-1 | A_{t_0} = j]$$

= P[after the first drawing urn 1 contains $j - 1$ black balls | urn 1 contained j black balls]

$$= P[B_1 W_2 | A_{t_0} = j]$$

$$= P[B_1 | A_{t_0} = j]\, P[W_2 | A_{t_0} = j] \text{ (since urns 1 and 2 are independent)}$$

$= \frac{j}{n} \cdot \frac{j}{n}$ (if there are j black balls in urn 1 then there are $n - j$ black balls in urn 2, so there are j white balls in urn 2)

$$= \left(\frac{j}{n}\right)^2$$

$$p_{jj} = P[A_{t_0+1} = j | A_{t_0} = j]$$

= P[after the first drawing urn 1 contains j black balls | urn 1 contained j black balls]

$$= P[W_1 W_2 \text{ or } B_1 B_2 | A_{t_0} = j]$$

$$= P[W_1 W_2 | A_{t_0} = j] + P[B_1 B_2 | A_{t_0} = j] \text{ (since } B_1 \text{ and } W_1 \text{ are mutually exclusive)}$$

$$= P[W_1 | A_{t_0} = j] P[W_2 | A_{t_0} = j] + P[B_1 | A_{t_0} = j] P[B_2 | A_{t_0} = j]$$

$$= \left(\frac{n-j}{n}\right)\left(\frac{j}{n}\right) + \left(\frac{j}{n}\right)\left(\frac{n-j}{n}\right) = 2\frac{(n-j)j}{n^2}$$

$$p_{jj+1} = P[A_{t_0+1} = j+1 | A_{t_0} = j] = (P[W_1 B_2 | A_{t_0} = j])$$

$$= P[W_1 | A_{t_0} = j] P[B_2 | A_{t_0} = j]$$

$$= \left(\frac{n-j}{n}\right)\left(\frac{n-j}{n}\right) = \left(\frac{n-j}{n}\right)$$

$p_{jk} = 0$ for $k > j+1$ or $k < j-1$ is obvious.

b. The one-step transition probability matrix is given as $P =$

	0	1	2	3	...	$n-3$	$n-2$	$n-1$	n
0	0	1	0	0	...	0	0	0	0
1	$\frac{1^2}{n^2}$	$\frac{2(n-1)(1)}{n^2}$	$\frac{(n-1)^2}{n^2}$	0	...	0	0	0	0
2	0	$\frac{2^2}{n^2}$	$\frac{2(n-2)(2)}{n^2}$	$\frac{(n-2)^2}{n^2}$	...	0	0	0	0
⋮	⋮	⋮	⋮	⋮	⋮	⋮	⋮	⋮	⋮
$n-2$	0	0	0	0	...	$\frac{(n-2)^2}{n^2}$	$\frac{2(2)(n-2)}{n^2}$	$\frac{2^2}{n^2}$	0
$n-1$	0	0	0	0	...	0	$\frac{(n-1)^2}{n^2}$	$\frac{2(1)(n-1)}{n^2}$	$\frac{1^2}{n^2}$
n	0	0	0	0	...	0	0	1	0

The stationary probabilities are found by solving for π in the recursive equation $\pi = \pi P$, The recursive equation gives us

$$\pi_0 = \pi_0 (0) + \pi_1 \left(\frac{1^2}{n^2}\right) + \pi_2 (0) + \ldots + \pi_n (0)$$

$$\pi_1 = \pi_0 (1) + \pi_1 \left(\frac{2 (n-1) (1)}{n^2} \right) + \pi_2 \left(\frac{2^2}{n^2}\right) + \pi_3 (0) + \ldots + \pi_n (0)$$

$$\vdots$$

$$\pi_n = \pi_0 (0) + \pi_1 (0) + \ldots + \pi_{n-1} \left(\frac{1^2}{n^2}\right) + \pi_n (0)$$

One can easily verify that

$$\pi_j = \pi_{j-1} \left(\frac{n - (j-1)}{j} \right)^2 \quad \text{for j=1, ..., n}$$

and hence that

$$\pi_j = \pi_0 \binom{n}{j}^2.$$

Next, from the fact that

$$\pi_0 + \pi_1 + \pi_2 + \ldots + \pi_n = 1$$

we have that

$$\pi_0 + \pi_0 \binom{n}{1}^2 + \pi_0 \binom{n}{2}^2 + \ldots + \pi_0 \binom{n}{n}^2 = 1$$

which gives us

$$\pi_0 \left(\sum_{j=0}^{n} \binom{n}{j}^2 \right) = 1$$

and hence

$$\pi_0 = \frac{1}{\sum_{j=0}^{n} \binom{n}{j}^2}$$

Note that $\binom{n+m}{r} = \binom{n}{0}\binom{m}{r} + \binom{n}{1}\binom{m}{r-1} + \ldots + \binom{n}{r}\binom{m}{0}$ since out of a group of n men and m women choosing a group of size $r\,(n \geq r, m \geq r)$ is comparable to choosing 0 men and r women or 1 man and r–1 women or ...

Now let $m = n$ and $r = n$. Then we have

$$\binom{2n}{n} = \binom{n}{0}\binom{n}{n} + \binom{n}{1}\binom{n}{n-1} + \ldots + \binom{n}{n}\binom{n}{0}$$

since $\binom{n}{k} = \binom{n}{n-k}$ then

$$\binom{2n}{n} = \binom{n}{0}\binom{n}{0} + \binom{n}{1}\binom{n}{1} + \ldots + \binom{n}{n}\binom{n}{n}$$

or that

$$\binom{2n}{n} = \sum_{j=0}^{n} \binom{n}{j}^2$$

Hence,

$$\pi_0 = \frac{1}{\binom{2n}{n}}$$

and thus

$$\pi_j = \frac{\binom{n}{j}^2}{\binom{2n}{n}} \text{ for j} = 0, 1, \ldots, n.$$

3.95

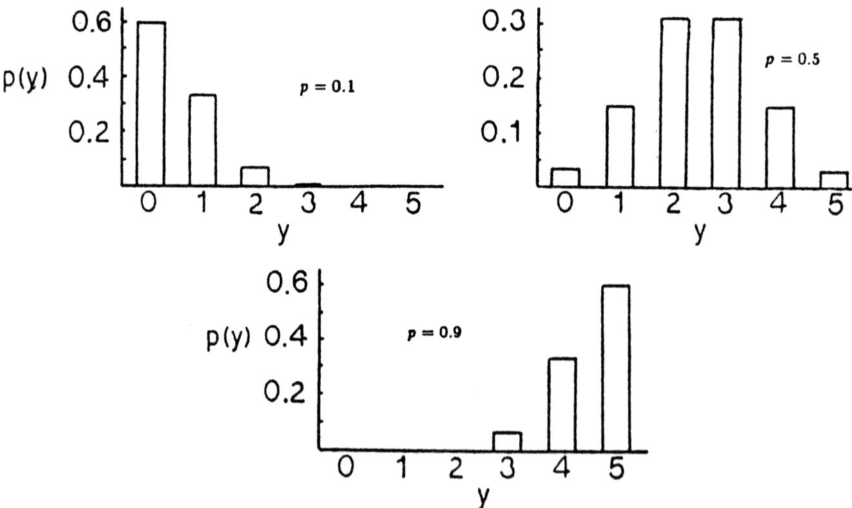

3.97 Let Y = number of radar sets out of the five that detect the plane. Then Y has a binomial distribution with parameters $n = 5$, $p = 0.9$,

$$P(Y=4) = \binom{5}{4}(0.9)^4(0.1)^1 = 0.32805$$

and

$$P(Y \geq 1) = 1 - P(Y=0) = 1 - \binom{5}{0}(0.9)^0(0.1)^5 = 0.99999.$$

3.99 $P(Y \leq a) = \binom{5}{0}p^0(1-p)^5 = (1-p)^5$

a. $(1-p)^5 = (1)^5 = 1$

b. $(1-p)^5 = (0.9)^5 = 0.5905$

c. $(1-p)^5 = (0.7)^5 = 0.1681$

d. $(1-p)^5 = (0.5)^5 = 0.03125$

e. $(1-p)^5 = (0)^5 = 0$

3.101 For $n = 5$, $a = 1$ we have

p	$P(Y \leq a)$
0.05	0.9774
0.10	0.9185
0.20	0.7373
0.30	0.5282
0.40	0.3370

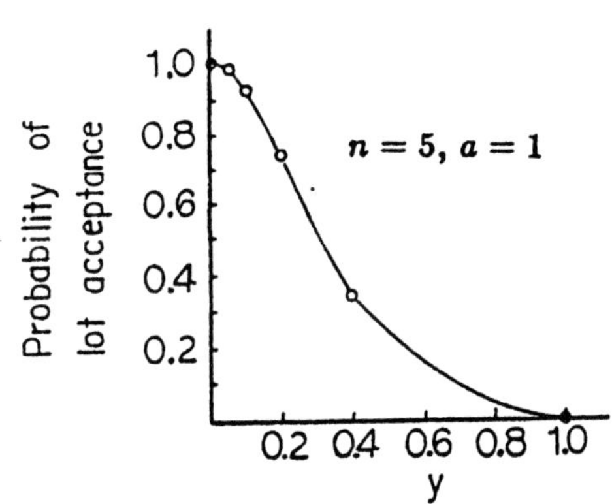

For $n = 25$, $a = 5$ we have

p	$P(Y \le a)$
0.05	0.9988
0.10	0.9666
0.20	0.6167
0.30	0.1935
0.40	0.0294

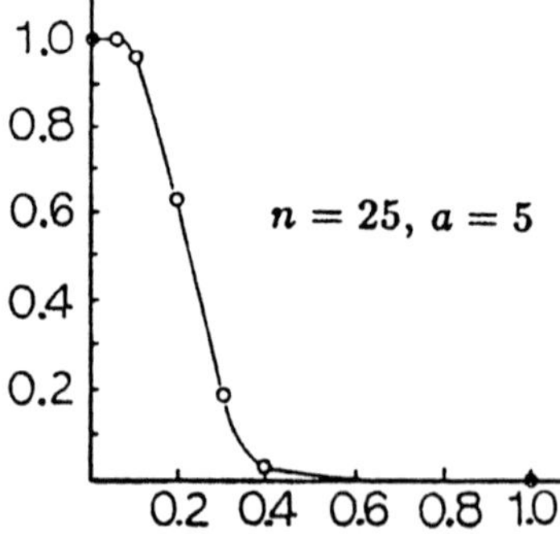

a. $n = 25$, $a = 5$

b. $n = 25$, $a = 5$

3.103 Let Y = number of colonies in a given dish. Then Y has a Poisson distribution with a mean of $\lambda = 12$.

a. Using Table 3, we have $P(Y \ge 10) = 1 - P(Y \le 9) = 1 - 0.242 = 0.758$.

b. $E(Y) = \lambda = 12$ and the standard deviation of $Y = \sqrt{V(Y)} = \sqrt{\lambda} = \sqrt{12} = 3.4641$.

c. By Tchebysheff's theorem, we have

$$P[E(Y) - 2\sqrt{V(Y)} < Y < E(Y) + 2\sqrt{V(Y)}] \ge 1 - \frac{1}{2^2} = 0.75$$

i.e., the desired interval is $[12 - 2(3.4641), 12 + 2(3.4641)] = (5.0718, 18.9282)$.

3.105 Here, Y has a Poisson distribution with mean $\lambda_p = 100(0.05) = 5$.

a. $E(Y) = \lambda p = 5$

b. Using Table 3, we have $P(Y = 0) = 0.007$.

c. Again using Table 3, we have $P(Y > 5) = 1 - P(Y \le 5) = 1 - 0.616 = 0.384$.

3.107 Let Y = number of left-turning vehicles out of n vehicles arriving while the light is red. Then Y has a binomial distribution with parameters $n = 5$, $p = 0.2$, and, using Table 2, we have $P(Y \le 3) = 0.993$. This may be computed directly as

$$\begin{aligned} P(Y \le 3) &= 1 - P(Y \ge 4) \\ &= 1 - P(Y = 4) - P(Y = 5) \\ &= 1 - \binom{5}{4}(0.2)^4(0.8)^1 - \binom{5}{5}(0.2)^5(0.8)^0 = 0.993 \end{aligned}$$

3.109 **a.** Using the Binomial theorem, we have

$$\sum p(y) = \sum_{y=0}^{n} \binom{n}{y} p^y (1-p)^{n-y} = [p + (1-p)] = 1^n = 1$$

b. $$\sum p(y) = \sum_{y=1}^{\infty} (1-p)^{y-1} p = p \sum_{y=1}^{\infty} (1-p)^{y-1} = p \sum_{y=0}^{\infty} (1-p)^y$$

$$= p\left(\frac{1}{1-(1-p)}\right) \quad \left(\text{since } \sum_{x=0}^{\infty} a^x = \frac{1}{1-a} \text{ for } a < 1\right)$$

$$= p\left(\frac{1}{p}\right) = 1$$

c. Noting that $\sum_{x=0}^{\infty} \frac{\lambda^x}{x!} = e^{\lambda}$, we have

$$\sum p(y) = \sum_{y=0}^{\infty} \frac{\lambda^y e^{-\lambda}}{y!} = e^{-\lambda} \sum_{y=0}^{\infty} \frac{\lambda^y}{y!} = e^{-\lambda} e^{\lambda} = 1$$

3.111 Let Y = total number of requests for welding units until the third brand A unit is used. Then Y has a negative binomial distribution with parameters $r = 3$, $p = 0.7$, and

$$P(Y=5) = \binom{5-1}{3-1} (0.7)^3 (0.3)^{(5-3)} = 6(0.7)^3(0.3)^2 = 0.18522.$$

3.113 Let Y = number of people that have to be interviewed before encountering a consumer who prefers brand A. Then Y has a geometric distribution with parameter $p = 0.6$, and

$$P(Y=5) = (1-p)^{5-1}\, p = (0.4)^4(0.6) = 0.01536$$

and

$$= 1 - p(1) - p(2) - p(3) - p(4)$$

$$= 1 - (0.6) - (0.4)(0.6) - (0.4)^2(0.6) - (0.4)^3(0.6)$$

$$= 1 - (0.6) - (0.24) - (0.096) - (0.0384)$$

$$= 0.0256.$$

3.115 Note that Y has a binomial distribution with parameters $n = 1000$, $p = 0.9$, so that

$$E(Y) = np = 1000(0.9) = 900, \text{ and}$$

$$V(Y) = np(1 - p) = 1000(0.9)(0.1) = 90.$$

Using Tchebysheff's theorem with $k = 2$, we have $P(\mu - 2\sigma < Y < \mu + 2\sigma) \geq 1 - \frac{1}{2^2}$; i.e.,

$$P(900 - 2\sqrt{90} < Y < 900 + 2\sqrt{90}) = P(881.026 < Y < 918.974) \geq 0.75.$$

3.117 **a.** Note that Y has a binomial distribution with parameters $n = 4$, $p = 1/3$, and distribution function

$$p(y) = \binom{4}{y}\left(\frac{1}{3}\right)^y\left(\frac{2}{3}\right)^{4-y}$$

y	$p(y)$
0	$\left(\frac{2}{3}\right)^4$
1	$4\left(\frac{1}{3}\right)\left(\frac{2}{3}\right)^3 = 2\left(\frac{2}{3}\right)^4$
2	$6\left(\frac{1}{3}\right)^2\left(\frac{2}{3}\right)^2 = \left(\frac{2}{3}\right)^3$
3	$4\left(\frac{1}{3}\right)^3\left(\frac{2}{3}\right)^1 = \frac{1}{3}\left(\frac{2}{3}\right)^3$
4	$\left(\frac{1}{3}\right)^4$

b. $P(Y \geq 3) = p(3) + p(4) = \frac{1}{3}\left(\frac{2}{3}\right)^3 + \left(\frac{1}{3}\right)^4 = \frac{1}{9}$

c. $E(Y) = np = 4\left(\frac{1}{3}\right) = \frac{4}{3}$

d. $V(Y) = np(1 - p) = 4\left(\frac{1}{3}\right)\left(\frac{2}{3}\right) = \frac{8}{9}$

3.119 **a.** Here, Y has a hypergeometric distribution with parameters $N = 100$, $n = 20$, $k = 40$, and

$$p(10) = \frac{\binom{40}{10}\binom{60}{10}}{\binom{100}{20}} = 0.1192$$

b. Here, Y, has a binomial distribution with parameters $n = 20$, $p = 0.40$, and using Table 2 we have

$$p(10) = F(10) - F(9) = 0.872 - 0.755 = 0.117$$

Thus it appears that N is large enough so that the binomial probability function is a good approximation to the hypergeometric probability function.

3.121 Let Y = number of items sold on a given day, P = daily profit, and X = number of items stocked. Note that for $Y \leq X$

$$P = 1.2Y - X,$$

$$E(P) = 1.2E(Y) - X$$

and

$$E(Y|X = 1) = 1$$

$$E(Y|X = 2) = 2$$

$$E(Y|X = 3) = 2p(2) + 3P(Y \geq 3) = 2(0.1) + 3(0.9) = 2.9$$

$$E(Y|X = 4) = 2p(2) + 3p(3) + 4p(4) = 2(0.1) + 3(0.4) + 4(0.5) = 3.4.$$

Hence

$$E(P|X = 1) = 1.2(1) - 1 = 0.2$$

$$E(P|X = 2) = 1.2(2) - 2 = 0.4$$

$$E(P|X = 3) = 1.2(2.9) - 3 = 0.48$$

$$E(P|X = 4) = 1.2(3.4) - 4 = 0.08.$$

Therefore, expected profit is maximized at $X = 3$.

3.123 Number of combinations $= 26 \cdot 26 \cdot 10 \cdot 10 \cdot 10 \cdot 10 = 6,760,000.$

E(winnings)

$$= \$100,000\left(\frac{1}{6,760,000}\right) + \$50,000\left(\frac{2}{6,760,000}\right) + \$1,000\left(\frac{10}{6,760,000}\right)$$

$$= \$0.031065.$$

Therefore, it appears that the expected value of the coupon is considerably less than the price of a stamp. However, one might also consider what the probability of winning would be if the coupon isn't mailed back.

CHAPTER 4
CONTINUOUS PROBABILITY DISTRIBUTIONS

4.1 **a.** Let Y = number of field plots out of ten that contain the insect. Y is a discrete random variable.

b. Let Y = number of defects in a given sampled section. Let X = number of sampled sections containing at least 5 defects. Both Y and X are discrete random variables.

c. Let Y = number of grains seen in a given cross section. Then Y is a discrete random variable.

d. Let A = area proportion covered by grains of a certain size. A is a continuous random variable.

4.3 **a.** $P(X>3) = \int_3^{\infty} f(x)\,dx = \int_3^6 \frac{3}{32}(8x-x^2-12)\,dx$

$$= \frac{3}{32}\left(4x^2 - \frac{x^3}{3} - 12x\right)\Bigg|_3^6 = \frac{27}{32} = 0.84375$$

b. $0.5 = P(X>b) = \int_b^6 \frac{3}{32}(8x-x^2-21)\,dx$

$$= \frac{3}{32}\left(4x^2 - \frac{x^3}{3} - 12x\right)\Bigg|_b^6 = 0 - \frac{3}{32}\left(4b^2 - \frac{b^3}{3} - 12b\right)$$

i.e., b is the solution to $0 = b^3 - 12b^2 + 36\,b - 16$; i.e., $b = 4$. Alternatively, note that the density of X is symmetric about $x = 4$, so $P(X > 4) = P(X \leq 4) = 1/2$.

4.5 **a.**

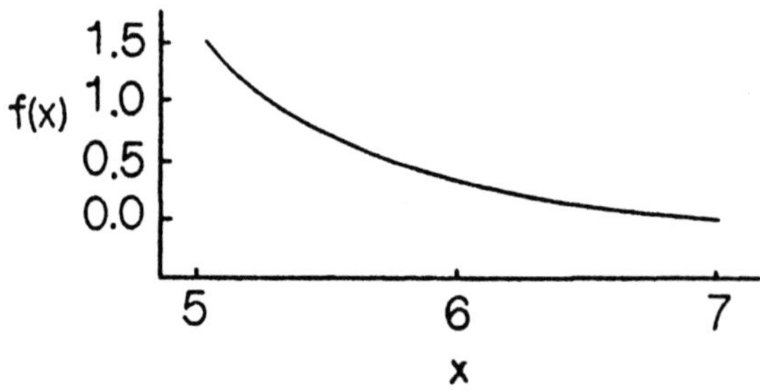

b. $$F(x)\begin{cases}0 & x<5\\ \int_5^x \frac{3}{8}(7-y)^2dy = \frac{3}{8}\int_{-2}^{(x-7)} w^2dw = \frac{w^3}{8}\Big|_{-2}^{x-7} = \frac{(x-7)^3}{8}+1 & 5\le x\le 7\\ 1 & x>7\end{cases}$$

c. $$P(X<6) = F(6) = \frac{(6-7)^3}{8}+1 = \frac{7}{8}$$

d. $$P(X<5.5|\,X<6) = \frac{P(X<5.5)}{P(X<6)} = \frac{\frac{(5.5-7)^3}{8}+1}{\frac{7}{8}} = \frac{37}{56}$$

4.7 **a.** $$P\left(X>\frac{1}{2}\right) = \int_{1/2}^{1} 2xdx = x^2\Big|_{1/2}^{1} = 1-\frac{1}{4} = \frac{3}{4}$$

b. $$P\left(X>\frac{1}{2}+X>\frac{1}{4}\right) = \frac{P\left(X>\frac{1}{2},X>\frac{1}{4}\right)}{P\left(X>\frac{1}{4}\right)} = \frac{P\left(\frac{1}{2}\right)}{P\left(X>\frac{1}{4}\right)} = \frac{\frac{3}{4}}{\int_{1/4}^{1} 2xdx} = \frac{4}{5}$$

c. $$P\left(X>\frac{1}{4}\,\Big|\,X>\frac{1}{2}\right) = \frac{P\left(X>\frac{1}{4},X>\frac{1}{2}\right)}{P\left(X>\frac{1}{2}\right)} = \frac{P\left(X>\frac{1}{2}\right)}{P\left(X>\frac{1}{2}\right)} = 1$$

d. $$F(x) = \begin{cases}0 & x<0\\ \int_0^x 2ydy = x^2 & 0\le x\le 1\\ 1 & x>1\end{cases}$$

Yes, $F(x)$ is continuous.

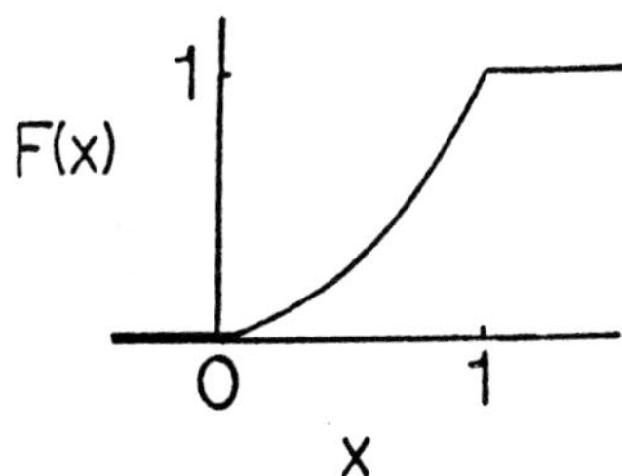

4.9 $E(X) = \int_{59}^{61} xf(x)\,dx = \int_{59}^{61} \frac{x}{2}dx = \frac{x^2}{4}\Big|_{59}^{61} = \frac{1}{4}[(61)^2 - (59)^2] = 60$

$$E(X^2) = \int_{59}^{61} x^2 f(x)\,dx = \int_{59}^{61} \frac{x^2}{2}dx = \frac{x^3}{6}\Big|_{59}^{61} = \frac{1}{6}[(61)^3 - (59)^3] = \frac{21602}{6}$$

$$V(X) = E(X^2) - [E(X)]^2 = \frac{21602}{6} - (60)^2 = \frac{1}{3}$$

4.11 $E(X) = \int_{-\infty}^{\infty} xf(x)\,dx = \frac{3}{32}\int_{2}^{6} x(x-2)(6-x)\,dx$

$$= -\frac{3}{32}\left(\frac{x^4}{4} - \frac{8x^3}{3} + 6x^2\right)\Big|_{2}^{6} = 4 \text{ hundred calories}$$

4.13 **a.** $E(X) = \frac{3}{8}\int_{5}^{7} x(7-x)^2 dx = \frac{3}{8}\left(\frac{49x^2}{2} - \frac{14x^3}{3} + \frac{x^4}{4}\right)\Big|_{5}^{7} = 5.5$

$$E(X^2) = \frac{3}{8}\int_{5}^{7} x^2(7-x)^2 dx = \frac{3}{8}\left(\frac{49x^3}{3} - \frac{7x^4}{4} + \frac{x^5}{5}\right)\Big|_{5}^{7} = 30.4$$

$$V(X) = E(X^2) - [E(X)]^2 = 30.4 - (5.5)^2 = 0.15$$

b. By Tchebysheff's theorem, with $k = 2$, we have

$$P[E(X) - 2\sqrt{V(X)} < X < E(X) + 2\sqrt{V(X)}] \geq 1 - \frac{1}{(2)^2} = 0.75$$

so the desired interval is $E(X) \pm 2\sqrt{V(X)} = (5.5 \pm 2\sqrt{0.15}) = (4.7254, 6.2746)$.

c. $P(X < 5.5) = \int_{-\infty}^{5.5} f(x)\,dx = \int_{5}^{5.5} \frac{3}{8}(7-x)^2 dx = \frac{3}{8}\left(49x - 7x^2 + \frac{x^3}{3}\right)\Big|_{5}^{5.5} = 0.5781$

We would expect to see about 58% of the pH measurements to be below 5.5.

4.15 **a.** $F(x) = \begin{cases} 0 & x < a \\ \int_{a}^{x} \frac{1}{b-a}dy = \frac{x-a}{b-a} & a \leq x \leq b \\ 1 & x > b \end{cases}$

b. $P(X > c) = \int_{c}^{b} \frac{1}{b-a}dx = \frac{b-c}{b-a}$

c. $P(X > d \mid X > c) = \frac{P(X > d, X > c)}{P(X > c)} = \frac{P(X > d)}{P(X > c)} = \frac{(b-d)/(b-a)}{(b-c)/(b-a)} = \frac{b-d}{b-c}$

4.17 Here, X has a uniform distribution with parameters $a = 0$, $b = 500$.

a. $P(X > 475) = \int_{475}^{500} \frac{1}{500-0} dx = \frac{500-475}{500} = \frac{1}{20}$

b. $P(X < 25) = \int_{0}^{25} \frac{1}{500-0} dx = \frac{20-0}{500} = \frac{1}{20}$

c. $P(X < 250) = \int_{0}^{250} \frac{1}{500-0} dx = \frac{250-0}{500} = \frac{1}{2}$

4.19 Let X = time in seconds that the call arrived. Then X has a uniform distribution with parameters $a = 0$, $b = 60$, and $P(X > 15) = \int_{15}^{60} \frac{1}{60-0} dx = \frac{60-15}{60} = \frac{3}{4}$.

4.21 Let X = hour of operation in which the defective board was produced. Since the number of defectives is Poisson, then given that one defective was produced, actual time of occurrence is equally likely in any small subinterval of time of a given size, and thus X has a uniform distribution with parameters $a = 0$, $b = 8$.

a. $P(X < 1) = \int_{0}^{1} \frac{1}{8} dx = \frac{1}{8}$

b. $P(X > 7) = \int_{7}^{8} \frac{1}{8} dx = \frac{1}{8}$

c. $P(4 < X \le 5 | X > 4) = \frac{P(4 < X \le 5, X > 4)}{P(X > 4)} = \frac{P(4 < X \le 5)}{P(X > 4)} = \frac{\int_{4}^{5} \frac{1}{8} dx}{\int_{4}^{8} \frac{1}{8} dx} = \frac{1}{4}$

4.23 Let X = measurement error. Then X has a uniform distribution with parameters $a = -0.02$, $b = 0.05$.

a. $P(-0.01 < X < 0.01) = \int_{-0.01}^{0.01} \frac{1}{0.05-(-0.02)} dx = \frac{2}{7}$

b. $E(X) = \frac{a+b}{2} = \frac{0.05+(-0.02)}{2} = 0.015$

$$V(X) = \frac{(b-a)^2}{12} = \frac{[0.05-(-0.02)]^2}{12} = \frac{49}{120,000} = 0.0004083.$$

4.25 Let X = time of arrival measured from the beginning of the 30-minute period. Since the number of arrivals is Poisson, the time of the arrival is equally likely in any subinterval of time of a given size in the 30 minutes, and thus X has a uniform distribution with parameters $a = 0$, $b = 30$, and

$$P(X > 25) = \int_{25}^{30} \frac{1}{30} dx = \frac{30-25}{30} = \frac{1}{6}.$$

4.27 Let X = stopping distance. Then X has a uniform distribution with parameters a and b.

a. $$P(X-a<b-X) = P\left(X< \frac{a+b}{2}\right) = \int_a^{\frac{a+b}{2}} \frac{1}{b-a}dx = \frac{\frac{a+b}{2}-a}{b-a} = \frac{1}{2}$$

b. $$P[X-a>3(b-X)] = P\left(X> \frac{3b+a}{4}\right) = \int_{\frac{3b+a}{4}}^{b} \frac{1}{b-a}dx = \frac{b-\frac{3b+a}{4}}{b-a} = \frac{1}{4}$$

4.29 Let X = cycle time. Then X has a uniform distribution with parameters $a = 50$, $b = 70$.

a. $$E(X) = \frac{a+b}{2} = \frac{50+70}{2} = 60$$

$$V(X) = \frac{(b-a)^2}{12} = \frac{(70-50)^2}{12} = \frac{100}{3}$$

b. Let T = number of trucks needed. Then we have $E(X)/T = 15$; i.e., $60/15 = 4$ trucks are needed.

4.31 Let X = magnitude of the next earthquake

a. $$P(X>3) = 1-F(3) = 1-(1-e^{-3/\theta}) = e^{-3/2.4} = e^{-5/4} = 0.2865$$

b. $$P(2<X<3) = F(3)-F(2) = 1-e^{-3/2.4}-(1-e^{-2/2.4}) = e^{-5/6}-e^{-5/4} = 0.1481$$

4.33 Let X = water demand. Then X has an exponential distribution with parameter $\theta = 100$.

a. $$P(X>200) = 1-F(200) = 1-(1-e^{-200/100}) = e^{-2} = 0.1353$$

b. Let k = maximum water-producing capacity. Then $0.01 = P(X > k) = 1 - F(k)$

$= 1 - (1 - e^{-k/100})$. Hence $k = -100 \ln 0.01 = 460.52$ cfs.

4.35 **a.** Note that since $\int_0^\infty x^{(n-1)} e^{-x/\theta} dx = \Gamma(n)\theta^n$, then for k integer valued we have

$$E(X^k) = \frac{1}{\theta}\int_0^\infty x^k e^{-x/\theta} dx = \frac{1}{\theta}\Gamma(k+1)\theta^{(k+1)} = \theta^k k!$$

So

$E(X) = (10)1! = 10$

$E(X^2) = (10)^2 2! = 200$

$E(X^3) = (10)^3 3! = 6000$

$E(X^4) = (10)^4 4! = 240{,}000$

and

$$E(C) = 100 + 40E(X) + 3E(X^2) = 100 + 40(10) + 3(200) = 1100$$

$$E(C^2) = E\{[100 + 40(X) + 3(X^2)]^2\}$$

$$= 10,000 + 8,000E(X) + 2,200E(X^2) + 240E(X^3) + 9E(X^4)$$

$$= 10,000 + 8,000(10) + 2,200(200) + 240(6,000) + 9(240,000)$$

$$= 4,130,000$$

$$V(C) = E(C^2) - [E(C)]^2 = 4,130,000 - (1,100)^2 = 2,920,000$$

b. $P(C > 2,000) = P(3X^2 + 40X + 100 > 2,000)$

$$= P(3X^2 + 40X - 1,900 > 0)$$

$$= P[(X - r_1)(X - r_2) > 0]$$

where $r_1 = \frac{10}{3}(-2 + \sqrt{61}) = 19.3675$, and $r_2 = \frac{10}{3}(-2 - \sqrt{61}) = -32.7$. Therefore

$$P(C > 2,000) = P(X - r_1 > 0, X - r_2 > 0) + P(X - r_1 < 0, X - r_2 < 0)$$

$$= P(X > r_1, X > r_2) + P(X < r_1, X < r_2)$$

$$= P(X > r_1) + P(X < r_2) = P(X > r_1)$$

$$= 1 - \left(1 - e^{-\frac{r_1}{10}}\right) = e^{-1.93675} = 0.1442.$$

4.37 Let X = tire life length. Then X has an exponential distribution with parameter $\theta = 30$.

a. $P(X > 30) = 1 - F(30) = 1 - (1 - e^{-30/30}) = e^{-1} = 0.3679$

b. Using the result of Exercise 4.30, we have $P(X > 30|X > 15) = P(X > 30 - 15) = 1 - F(15)$

$= 1 - (1 - e^{-15/30}) = e^{-1/2} = 0.6065.$

4.39 If the number of breakdowns has a Poisson distribution with parameter $\lambda = 0.5$, then the waiting time between breakdowns, X, has an exponential distribution with parameter $\theta = \frac{1}{\lambda} = \frac{1}{0.5} = 2.$

a. $P(X > 1) = 1 - F(1) = 1 - (1 - e^{-1/2}) = e^{-1/2} = 0.6065$

b. $P\left(X > \frac{1}{2}\right) = 1 - (1 - e^{-0.5/2}) = e^{-1/4} = 0.7788$

c. No, because of the result of Exercise 4.30, the "memoryless" property of the exponential distribution.

4.41 Let X = the weekly rainfall totals.

a. $P(X > 2) = 1 - F(2) = 1 - (1 - e^{-2/1.6}) = e^{-5/4} = 0.2865$

b. Let Y = number of weeks out of the next two in which rainfall doesn't exceed 2 inches. Then Y has a binomial distribution with parameters $n = 2$, $p = P(X \le 2) = F(2) = 1 - e^{-5/4}$. Then

$$P(Y = 2) = \binom{2}{2}(1 - e^{-5/4})^2 (e^{-5/4})^0 = (1 - e^{-5/4})^2 = 0.5091.$$

4.43 Let X = repair time.

a. $P(X < 10) = F(10) = 1 - e^{-10/22} = 0.3653$

b. $P(30 < X < 60) = F(60) - F(30) = 1 - e^{-60/22} - (1 - e^{-30/22}) = e^{-15/11} - e^{-30/11} = 0.1903$

c. Let k = the desired constant such that $P(X > k) = 1 - F(k) = 1 - (1 - e^{-k/22}) = e^{-k/22} = 0.10$. Solving for k, we have $k = -22 \ln(0.10) = 50.66$ minutes.

4.45 **a.** Let X = summer rainfall total.

$$E(X) = \alpha\beta = 1.6(2) = 3.2$$

$$V(X) = \alpha\beta^2 = 1.6(2)^2 = 6.4$$

b. Using Tchebysheff's theorem with $k = 2$, we have $0.75 = 1 - \frac{1}{2^2} \le P[E(X) - 2\sqrt{V(X)} < X < E(X) + 2\sqrt{V(X)}] = P(3.2 - 2\sqrt{6.4} < X < 3.2 + 2\sqrt{6.4}) = P(-1.86 < X < 8.26)$. Since rainfall totals are nonnegative, we have the interval (0, 8.26).

4.47 For k integer valued, note that

$$E(Y^k) = \int_0^\infty \frac{1}{\Gamma(\alpha)\beta^\alpha} y^{(\alpha+k-1)} e^{-y/\beta} dy$$

$$= \frac{\beta^k \Gamma(\alpha+k)}{\Gamma(\alpha)} \int_0^\infty \frac{1}{\Gamma(\alpha+k)\beta^{\alpha+k}} y^{(\alpha+k-1)} e^{-y/\beta} dy = \frac{\beta^k \Gamma(\alpha+k)}{\Gamma(a)}$$

a. $E(L) = 30E(Y) + 2E(Y^2) = 30\alpha\beta + \frac{2\beta^2\Gamma(\alpha+2)}{\Gamma(\alpha)} = 30\alpha\beta + \frac{2\beta^2(\alpha+1)!}{(\alpha-1)!}$

$$= 30(3)(2) + \frac{2(2)^2(24)}{2} = 276$$

$$V(L) = E(L^2) - [E(L)]^2 = E[(30Y + 2Y^2)^2] - (276)^2$$

$$= 900E(Y^2) + 120E(Y^3) + 4E(Y^4) - (276)^2$$

$$= 900(2)^2\frac{\Gamma(5)}{\Gamma(3)} + 120(2)^3\frac{\Gamma(6)}{\Gamma(3)} + 4(2)^4\frac{\Gamma(7)}{\Gamma(3)} - (276)^2$$

$$= 900(4)\frac{24}{2} + 120(8)\frac{120}{2} + 4(16)\frac{720}{2} - (276)^2 = 47,664$$

b. Using Tchebysheff's theorem, we want k such that $1 - \frac{1}{k^2} = 0.89$; i.e., $k \approx 3$. Then the desired interval is

$[E(L) - 3\sqrt{V(L)}\ ,\ E(L) + 3\sqrt{V(L)}] = (276 - 3\sqrt{47,644}\ , 276 + 3\sqrt{47,664})$

$= (-378.963, 930.963)$. Since L is nonnegative, the interval is $(0, 930.963)$.

4.49 Let X_i = time to completion of the given task, $i = 1, 2$. Then X_i has a Gamma distribution with parameters $\alpha = 1, B = 10$, and $Y = X_1 + X_2$ has a Gamma distribution with parameters $\alpha = 2(1) = 2$, $\beta = 10$.

a. $E(Y) = \alpha\beta = 2(10) = 20$, and $V(Y) = \alpha\beta^2 = 2(10)^2 = 200$

b. Let A = average time to completion of the two tasks $= \frac{X_1 + X_2}{2} = \frac{Y}{2}$. Then $E(A) = \frac{1}{2}E(Y) = \frac{1}{2}(20) = 10$ and $V(A) = \left(\frac{1}{2}\right)^2 V(Y) = \frac{1}{4}(200) = 50$.

4.51 **a.** Let Y = maximum river flow. Then Y has a Gamma distribution with parameters $\alpha = 1.6$, $\beta = 150$. Therefore we have

$$E(Y) = \alpha\beta = 1.6(150) = 240$$

$$V(Y) = \alpha\beta^2 = 1.6(150)^2 = 36,000$$

$$\text{Std dev(Y)} = \sqrt{V(Y)} = \sqrt{36,000} = 189.74.$$

b. Using Tchebysheff's theorem, we want k such that $1 - \frac{1}{k^2} = \frac{8}{9}$; i.e., $k = 3$. Then the desired interval is $[E(Y) - 3\sqrt{V(Y)},\ E(Y) + 3\sqrt{V(Y)}] = (240 - 3\sqrt{36,000}, 240 + 3\sqrt{36,000})$ $= (-329.21, 809.21)$. Since Y is nonnegative, the interval is $(0, 809.21)$.

4.53 Since service times have a Gamma distribution with parameters $\alpha = 1$, $\beta = 3.2$, then the service time for three waiting customers, Y, has a Gamma distribution with parameters $\alpha = 3(1) = 3$, $\beta = 3.2$, and

$$E(Y) = \alpha\beta = 3(3.2) = 9.6$$

$$V(Y) = \alpha\beta^2 = 3(3.2)^2 = 30.72$$

$$f(y) = \begin{cases} \dfrac{1}{3.2^3\Gamma(3)} y^2 e^{-\frac{y}{3.2}} = \dfrac{1}{65.536} y^2 e^{-\frac{y}{3.2}} & y > 0 \\ 0 & y \le 0 \end{cases}.$$

4.55 **a.** $P(0 \le Z \le 1.2) = 0.3849$

b. $P(-0.9 \le Z \le 0) = P(0 \le Z \le 0.9) = 0.3159$

c. $P(0.3 \le Z \le 1.56) = P(0 \le Z \le 1.56) - P(0 \le Z \le 0.3) = 0.4406 - 0.1179 = 0.3227$

d. $P(-0.2 \le Z \le 0.2) = 2P(0 \le Z \le 0.2) = 2(0.0793) = 0.1586$

e. $P(-2.00 \le Z \le 1.56) = P(0 \le Z \le 2.00) + P(0 \le Z \le 1.56) = 0.4772 + 0.4406 = 0.9178$

4.57 Let X = amount spent on maintenance and repairs. Then X has a normal distribution with parameters $\mu = 400$, $\sigma = 20$ and

$$P(X > 450) = P\left(\frac{X - 400}{20} > \frac{450 - 400}{20}\right)$$

$$= P(Z > 2.5) = 0.5 - 0.4938 = 0.0062.$$

4.59 Let X = diameter. Then X has a normal distribution with parameters $\mu = 1.005$, $\sigma = 0.01$, and

$$P(X < 0.98) + P(X > 1.02) = P\left(Z < \frac{0.98 - 1.005}{0.01}\right) + P\left(Z > \frac{1.02 - 1.005}{0.01}\right)$$

$$= P(Z < -2.5) + P(Z > 1.5)$$

$$= (0.5 - 0.4938) + (0.5 - 0.4332) = 0.0730.$$

4.61 Let X = resistances of wires produced by Company A. Then X has a normal distribution with parameters $\mu = 0.13$, $\sigma = 0.005$.

a. $$P(0.12 < X < 0.14) = P\left(\frac{0.12 - 0.13}{0.005} < \frac{X - 0.13}{0.005} < \frac{0.14 - 0.13}{0.005}\right)$$

$$= P(-2 < Z < 2) = 2P(0 < Z < 2)$$

$$= 2(0.4772) = 0.9544$$

b. Let Y = number of wires of a sample of four from Company A that meet specifications. Then Y has a binomial distribution with parameters $n = 4$, $p = 0.9544$, and

$$P(Y = 4) = \binom{4}{4}(0.9544)^4(1 - 0.9544)^0 = 0.8297.$$

4.63 $P(|X| > 5) = P(X < -5) + P(X > 5)$

$$= P\left(Z < \frac{-5-0}{10}\right) + P\left(Z > \frac{5-0}{10}\right)$$

$$= 2P(Z > 0.5) = 2(0.5 - 0.1915) = 0.6170$$

$P(|X| > 10) = P(X < -10) + P(X > 10)$

$$= P\left(Z < \frac{-10-0}{10}\right) + P\left(Z > \frac{10-0}{10}\right)$$

$$= 2P(Z > 1) = 2(0.5 - 0.3413) = 0.3174$$

4.65 Let X = monthly sick-leave time. Then X has a normal distribution with parameters $\mu = 200$, $\sigma = 20$.

a. $P(X < 150) = P\left(\frac{X-200}{20} < \frac{150-200}{20}\right) = P(Z < -2.5) = 0.5 - 0.4938 = 0.0062$

b. Let x_0 = desired time budgeted. Then $P(X > x_0) = P\left(Z > \frac{x^0 - 200}{20}\right) \underset{=}{\text{set}} 0.1$. Note that $P(Z > 1.28) = 0.1$ so we have $\frac{x_0 - 200}{20} = 1.28$; i.e., $x_0 = 225.6$ hours.

4.67 Let X = amount of fill per box. Then X has a normal distribution with parameters μ, $\sigma = 1$, and $P(X > 16) = P\left(Z > \frac{16-\mu}{1}\right) \underset{=}{\text{set}} 0.01$. Note that $P(Z > 2.33) = 0.01$, so we have $\frac{16-\mu}{1} = 2.33$; i.e., $\mu = 13.67$ ounces.

4.69 **a.** Yes, it does appear that the total points can be modeled by a normal distribution.

b. According to the empirical rule, 68% of the data should lie one standard deviation above and below the mean and 95% of the data should lie within two standard deviations above and below the mean. Hence, consider the interval $(\bar{x} - s,\ \bar{x} + s) = (143 - 26,\ 143 + 26) = (117, 169)$. Notice that more than 77% of the games had total scores within (117, 169). Now consider the interval $(\bar{x} - 2s,\ \bar{x} + 2s) = (143 - 2(26),\ 143 + 2(26)) = (91, 195)$. Notice that less than 5% of the total scores fall outside of this region.

c. No and no. A score of 200 is greater than two standard deviations away from the mean. Such a score should occur less than 2.5% of the time, according to the empirical rule. A score of 250 is greater than three standard deviations away from the mean, making it even less likely to occur.

d. About 4 games.

4.71 **a.** Discussion question. (Q-Q plot is in the Appendix.)

b. $\bar{x} \approx 64 \quad s \approx 9.5$

4.73 **a.** $1 = \int_0^1 kx^3(1-x)^2 dx = k\int_0^1 x^{(4-1)}(1-x)^{(3-1)}dx$

$$= k\frac{\Gamma(4)\Gamma(3)}{\Gamma(7)} = k\frac{6(2)}{720} = k\frac{1}{60}$$

i.e., $k = \dfrac{\Gamma(7)}{\Gamma(4)\Gamma(3)} = 60$, and X has a Beta distribution with parameters $\alpha = 4$, $\beta = 3$.

b. $E(X) = \dfrac{\alpha}{\alpha+\beta} = \dfrac{4}{7}$

$$V(X) = \frac{\alpha\beta}{(\alpha+\beta)^2(\alpha+\beta+1)} = \frac{4(3)}{(7)^2(8)} = \frac{3}{98}$$

4.75 Note that

$$E(X^k) = \int_0^1 x^k \frac{\Gamma(\alpha+\beta)}{\Gamma(\alpha)\Gamma(\beta)} x^{\alpha-1}(1-x)^{\beta-1}dx$$

$$= \int_0^1 \frac{\Gamma(\alpha+\beta)}{\Gamma(\alpha)\Gamma(\beta)} x^{\alpha+k-1}(1-x)^{\beta-1}dx$$

$$= \frac{\Gamma(\alpha+\beta)\Gamma(\alpha+k)}{\Gamma(\alpha)\Gamma(\alpha+k+\beta)}\int_0^1 \frac{\Gamma(\alpha+\beta+k)}{\Gamma(\alpha+k)\Gamma(\beta)} x^{\alpha+k-1}(1-x)^{\beta-1}dx$$

$$= \frac{\Gamma(\alpha+\beta)\Gamma(\alpha+k)}{\Gamma(\alpha)\Gamma(\alpha+k+\beta)}$$

a. $E(C) = 10 + 20E(X) + 4E(X^2)$

$$= 10 + (20)\frac{\alpha}{\alpha+\beta} + (4)\frac{\Gamma(\alpha+\beta)\Gamma(\alpha+2)}{\Gamma(\alpha)\Gamma(\alpha+2+\beta)}$$

$$= 10 + (20)\frac{1}{3} + (4)\frac{\Gamma(3)\Gamma(3)}{\Gamma(1)\Gamma(5)}$$

$$= 10 + \frac{20}{3} + (4)\frac{(2)(2)}{24} = \frac{52}{3} = 17.33$$

$$E(C^2) = E[10 + 20X + 4(X^2)]^2$$

$$= 100 + 400E(X) + 480E(X^2) + 160E(X^3) + 16E(X^4)$$

$$= 100 + (400)\frac{1}{3} + (480)\frac{1}{6} + (160)\frac{\Gamma(3)\Gamma(4)}{\Gamma(1)\Gamma(6)} + (16)\frac{\Gamma(3)\Gamma(5)}{\Gamma(1)\Gamma(7)}$$

$$= 100 + \frac{400}{3} + \frac{480}{6} + \frac{160}{10} + \frac{16}{15} = \frac{1652}{5}$$

Then we have

$$V(C) = E(C^2) - [E(C)]^2 = \frac{1,652}{5} - \left(\frac{52}{3}\right)^2 = \frac{1,348}{45} = 29.96.$$

b. Using Tchebysheff's theorem with $k = 2$, we have $1 - \frac{1}{k^2} = \frac{3}{4}$, so the desired interval is

$$[E(C) - 2\sqrt{V(C)}, E(C) + 2\sqrt{V(C)}] = \left(\frac{52}{3} - 2\sqrt{\frac{1,348}{45}}, \frac{52}{3} + 2\sqrt{\frac{1,348}{45}}\right)$$

$$= (6.387, 28.280)$$

4.77 $E(X) = \frac{\alpha}{\alpha+\beta} = \frac{4}{4+2} = \frac{2}{3}$. The angle corresponding to $E(X)$ is $360°\ E(X) = 360°(2/3) =$ $240°$.

4.79 **a.** $E(X) = \frac{\alpha}{\alpha+\beta} = \frac{3}{6} = \frac{1}{2}$

$$V(X) = \frac{\alpha\beta}{(\alpha+\beta)^2(\alpha+\beta+1)} = \frac{9}{(3+3)^2(3+3+1)} = \frac{1}{28}$$

b. $E(X) = \frac{\alpha}{\alpha+\beta} = \frac{2}{4} = \frac{1}{2}$

$$V(X) = \frac{\alpha\beta}{(\alpha+\beta)^2(\alpha+\beta+1)} = \frac{4}{(2+2)^2(2+2+1)} = \frac{1}{20}$$

c. $E(X) = \frac{\alpha}{\alpha+\beta} = \frac{1}{2}$

$$V(X) = \frac{\alpha\beta}{(\alpha+\beta)^2(\alpha+\beta+1)} = \frac{4}{(1+1)^2(1+1+1)} = \frac{1}{12}$$

d. Case (a) exemplifies the best blending since it has the smallest variance.

4.81 Let X = fatigue life. Then X has a Weibull distribution with parameters $\gamma = 2$, $\theta = 4$.

a. $P(Y<2) = F(2) = 1 - e^{-\frac{2^2}{4}} = 1 - e^{-1} = 0.6321$

b. $E(X) = \theta^{1/\gamma}\Gamma\left(1+\frac{1}{\gamma}\right) = 4^{1/2}\Gamma\left(1+\frac{1}{2}\right) = 2\left(\frac{1}{2}\right)\Gamma\left(\frac{1}{2}\right) = \Gamma\left(\frac{1}{2}\right) = \sqrt{\pi}$

4.83 Let X = the time necessary to achieve proper blending of copper powders. Then X has a Weibull distribution with parameters $\gamma = 1.1$, $\theta = 2$, and

$$P(X<2) = F(2) = 1 - e^{-\frac{(2)^{1.1}}{2}} = 0.6576.$$

4.85 Let X = steel-beam yield strength. Then X has a Weibull distribution with parameters $\gamma = 2$, $\theta = 3{,}600$. Let Y = number of beams out of two that have yield strengths greater than 70,000 psi. Then Y has a binomial distribution with parameters $n = 2$, $p = P(X > 70) = 1 - F(70) = 1 - (1 - e^{-\frac{(70)^2}{3600}}) = e^{-49/36}$, and

$$P(Y = 2) = \binom{2}{2}\left(e^{-\frac{49}{36}}\right)^2\left(1 - e^{-\frac{49}{36}}\right)^0 = 0.06573.$$

4.87 Let X = resistor life length in thousands of hours. Then X has a Weibull distribution with parameters $\gamma = 2$, $\theta = 10$.

a. $P(X > 5) = 1 - F(5) = 1 - \left(1 - e^{-\frac{(5)^2}{10}}\right) = e^{-2.5} = 0.08208$

b. Let Y = number of resistors out of three that fail before 5,000 hours. Then Y has a binomial distribution with parameters $n = 3$, $p = P(X < 5) = F(5) = 1 - e^{-2.5}$, and

$$P(Y = 1) = \binom{3}{1}(1 - e^{-2.5})^1(e^{-2.5})^2 = 0.01855.$$

c. $E(X) = \theta^{1/\gamma}\Gamma\left(1 + \frac{1}{\gamma}\right) = 10^{1/2}\Gamma\left(1 + \frac{1}{2}\right) = 10^{1/2}\left(\frac{1}{2}\right)\Gamma\left(\frac{1}{2}\right) = \frac{\sqrt{10\pi}}{2} = 2.8025$

$$V(X) = \theta^{2/\gamma}\{\Gamma\left(1 + \frac{2}{\gamma}\right) - \left[\Gamma\left(1 + \frac{1}{\gamma}\right)\right]^2\} = 10^{2/2}\{\Gamma\left(1 + \frac{2}{2}\right) - \left[\Gamma\left(1 + \frac{1}{2}\right)\right]^2\}$$

$$= 10\{1 - \left[\frac{1}{2}\Gamma\left(\frac{1}{2}\right)\right]^2\} = 2.1460 = 10\left(1 - \frac{\pi}{4}\right) = 2.1460$$

4.89 Let X = maximum wind gust velocity. Then X has a Weibull distribution with parameters $\gamma = 2$, $\theta = 400$. Let k = the desired velocity such that $0.01 = P(X > k) = 1 - F(k) = 1 - (1 - e^{-\frac{(k)^2}{400}}) = e^{-\frac{(k)^2}{400}}$. Then, solving for k, we have $k = \sqrt{400[-\ln(0.01)]} = 42.9293$ feet per second.

4.91 $R(t_2 + t_1 | t_1) = P(X > t_2 + t_1 | X > t_1)$

$$= \frac{P(X > t_2 + t_1, X > t_1)}{P(X > t_1)} = \frac{P(X > t_2 + t_1)}{P(X > t_1)} = \frac{e^{-(t_2 + t_1)/\theta}}{e^{-t_1/\theta}} = e^{-t_2/\theta}, t_2 > 0 = R(t_2)$$

4.93 For system (a), the reliability would be

$$R_s^a(t) = 1 - [1 - (R(t))^2][1 - R(t)]$$

$$= R(t) + (R(t))^2 - (R(t))^3.$$

For system (b), the reliability would be

$$R_s^b(t) = R(t)[1 - [1 - R(t)]^2]$$

$$= 2(R(t))^2 - (R(t))^3.$$

Consider $R_s^a(t) - R_s^b(t) = R(t) + (R(t))^2 - (R(t))^3 - 2(R(t))^2 + (R(t))^3 = R(t) - (R(t))^2$ > 0.

So system (a) is more reliable.

4.95 $M(t) = E(e^{tX})$

$$= \int_0^\infty e^{tx} \frac{1}{\Gamma(\alpha)\beta^\alpha} x^{(\alpha-1)} e^{-\frac{x}{\beta}} dx$$

$$= \int_0^\infty \frac{1}{\Gamma(\alpha)\beta^\alpha} x^{(\alpha-1)} e^{-x(1/\beta - t)} dx \qquad \left(t < \frac{1}{\beta}\right)$$

$$= (1-\beta t)^{-\alpha} \int_0^\infty \frac{1}{\Gamma(\alpha)\left(\frac{\beta}{1-\beta t}\right)^\alpha} x^{(\alpha-1)} e^{-x/\left(\frac{\beta}{1-\beta t}\right)} dx$$

$$= (1-\beta t)^{-\alpha}$$

4.97 $M_Z(t) = E(e^{tZ})$

$$= \int_{-\infty}^\infty e^{tZ} \frac{1}{\sqrt{2\pi}} e^{-\frac{x^2}{2}} dz = e^{t^2/2} \int_{-\infty}^\infty \frac{1}{\sqrt{2\pi}} e^{-\frac{1}{2}(x-t)^2} dz$$

Since the integrand is a normal density function with parameters $\mu = t$, $\sigma = 1$, we have $M_Z(t) = e^{t^2/2}$.

4.99 Let Y = amount of food bought by the grocer. Next, let

$$Z = \begin{cases} 10X - 6Y & \text{if } 0 \le X < Y \\ 4Y & \text{if } Y < X \le 1 \end{cases}$$

Then Z is the profit or loss realized by the grocer. To maximize the profit, the expected value of Z based on Y is needed. Hence,

$$E(Z) = \int_0^y 10x - 6y dx + \int_y^1 4y dx = 5x^2 - 6yx|_0^y + 4yx|_y^1 = 4y - 5y^2$$

To find the maximum of $E(Z)$, we differentiate with respect to y and set it equal to zero and solve. Thus,

$$\frac{d}{dy}E(Z) = \frac{d}{dy}(4y - 5y^2) = 4 - 10y = 0$$

So $y = .4$ pounds will maximize the grocer's expected profit.

4.101 **a.** Exponential pieces with jumps at multiples of 3.

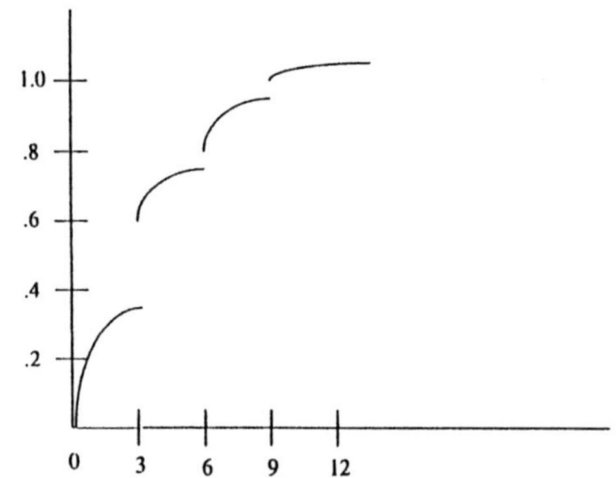

b. $P(X \le 6) = 1 - \frac{2}{3}e^{-2} - \frac{1}{3}e^{-2} = 1 - e^{-2}$

c. $P(X > 4) = \frac{2}{3}e^{-4/3} + \frac{1}{3}e^{-1}$

d. $F(x) = \frac{1}{3}[1 - e^{-[x/3]}] + \frac{2}{3}[1 - e^{-x/3}]$

$F_1(x) = 1 - e^{-[x/3]}$, $x = 0, 3, 6, 9, \ldots$ is a discrete d.f.

$F_2(x) = 1 - e^{-x/3}$ is a continuous d.f. (exponential) for $x > 0$.

e. $E(X_1) = 3[1 - e^{-1}] + 6[e^{-1} - e^{-2}] + 9[e^{-2} - e^{-3}] + \ldots$

$= 3[1 + e^{-1} + e^{-2} \ldots] = 3\left[\frac{1}{1 - e^{-1}}\right]$

$E(X_2) = 3$

Therefore,

$$E(X) = \frac{1}{3}E(X_1) + \frac{2}{3}E(X_2) = \frac{1}{3} \cdot \frac{3}{1-e^{-1}} + \frac{2}{3} \cdot 3 = \frac{1}{1-e^{-1}} + 2$$

4.103 **a.** $1 = \int_0^2 (cy^2 + y)\,dy = \left(\frac{cy^3}{3} + \frac{y^2}{2}\right)\Bigg|_0^2 = c\left(\frac{8}{3}\right) + 2$

i.e., $c = -3/8$.

b. $F(y) = \begin{cases} 0 & y < 0 \\ \int_0^y \left(x - \frac{3x^2}{8}\right)dx = \frac{y^2}{2} - \frac{y^3}{8} & 0 \le y \le 2 \\ 1 & y > 2 \end{cases}$

c.

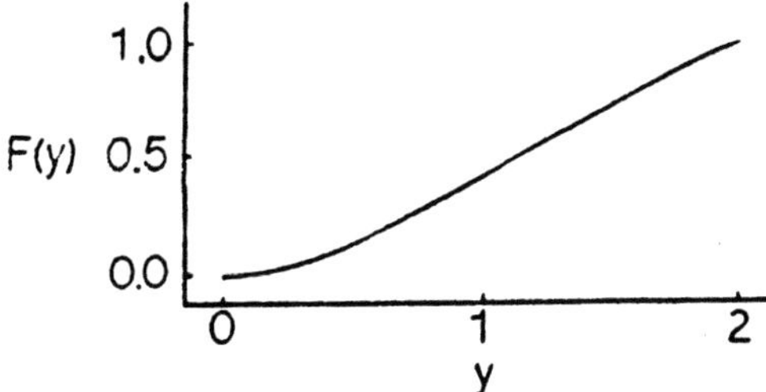

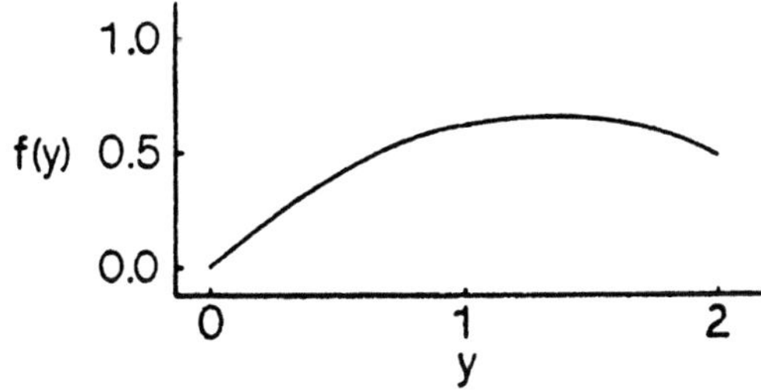

d. $F(-1) = 0$

$F(0) = 0$

$F(1) = 1/2 - 1/8 = 3/8$

e. $P\left(0 < Y < \frac{1}{2}\right) = F\left(\frac{1}{2}\right) - F(0) = \frac{(1/2)^2}{2} - \frac{(1/2)^3}{8} = \frac{1}{8} - \frac{1}{64} = \frac{7}{64}$

f. $$E(Y) = \int_0^2 y\left(y - \frac{3y^2}{9}\right)dy = \left(\frac{y^3}{3} - \frac{3y^4}{32}\right)\Bigg|_0^2 = \left(\frac{8}{3} - \frac{3}{2} = \frac{7}{6}\right)$$

$$E(Y^2) = \int_0^2 y^2\left(y - \frac{3y^2}{8}\right)dy = \left(\frac{y^4}{4} - \frac{3y^5}{40}\right)\Bigg|_0^2 = 4 - \frac{12}{5} = \frac{8}{5}$$

$$V(Y) = E(Y^2) - [E(Y)]^2 = \frac{8}{5} - \left(\frac{7}{6}\right)^2 = \frac{43}{180} = 0.2389$$

4.105 Let X = student GPA. $P(X > 3) = P\left(\frac{X - 2.4}{0.5} > \frac{3 - 2.4}{0.5}\right) = P(Z > 1.2) = 0.5 - 0.3849 = 0.1151$

4.107 Let Y = number of students out of three that possess a GPA in excess of 3.0. Then Y has a binomial distribution with parameters $n = 3$, $p = P(X > 3) = 0.1151$, and

$$P(Y = 3) = \binom{3}{3}(0.1151)^3(1 - 0.1151)^0 = (0.1151)^3 = 0.001525.$$

4.109 Let Y = number of defective bearings out of a sample of five. Then Y has a binomial distribution with parameters $n = 5$, $p = P(\text{bearing is scrap}) = 0.073$ (from Exercise 4.108), and

$$P(Y \geq 1) = 1 - P(Y = 0) = 1 - \binom{5}{0}(0.073)^0(1 - 0.073)^5$$

$$= 1 - (0.927)^5 = 0.3155$$

4.111 $$E(X^k) = \int_0^1 x^k \frac{\Gamma(\alpha+\beta)}{\Gamma(\alpha)\Gamma(\beta)} x^{\alpha-1}(1-x)^{\beta-1}dx = \frac{\Gamma(\alpha+\beta)}{\Gamma(\alpha)\Gamma(\beta)}\int_0^1 x^{\alpha+k-1}(1-x)^{\beta-1}dx$$

$$= \frac{\Gamma(\alpha+\beta)\Gamma(\alpha+k)}{\Gamma(\alpha+\beta+k)\Gamma(\alpha)}\int_0^1 \frac{\Gamma(\alpha+\beta+k)}{\Gamma(\alpha+k)\Gamma(\beta)} x^{\alpha+k-1}(1-x)^{\beta-1}dx$$

$$= \frac{\Gamma(\alpha+\beta)\Gamma(\alpha+k)}{\Gamma(\alpha+\beta+k)\Gamma(\alpha)}$$

Therefore,

$$E(X) = \frac{\Gamma(\alpha+\beta)\Gamma(\alpha+1)}{\Gamma(\alpha+\beta+1)\Gamma(\alpha)} = \frac{\alpha}{\alpha+\beta}$$

$$E(X^2) = \frac{\Gamma(\alpha+\beta)\Gamma(\alpha+2)}{\Gamma(\alpha+\beta+2)\Gamma(\alpha)} = \frac{(\alpha+1)\alpha}{(\alpha+\beta+1)(\alpha+\beta)}$$

$$V(X) = E(X^2) - [E(X)]^2 = \frac{(\alpha+1)\alpha}{(\alpha+\beta+1)(\alpha+\beta)} - \left(\frac{\alpha}{\alpha+\beta}\right)^2$$

$$= \frac{(\alpha+\beta)(\alpha+1)\alpha - \alpha^2(\alpha+\beta+1)}{(\alpha+\beta+1)(\alpha+\beta)^2} = \frac{\alpha\beta}{(\alpha+\beta+1)(\alpha+\beta)^2}$$

4.113 Let X = gap time. Then X has an exponential distribution with parameter $\theta = 10$, which is equivalent to a Gamma distribution with parameters $\alpha = 1, \beta = \theta = 10$.

a. $P(X \le 60) = F(60) = 1 - e^{-\frac{60}{10}} = 1 - e^{-6} = 0.9975$

b. Assuming independent gap times, Y, the sum of the next four gap times, has a Gamma distribution with parameters $\alpha = 4, \beta = 10$; i.e.,

$$f(y) = \begin{cases} \dfrac{1}{\Gamma(4)\,(10)^4} y^{(4-1)} e^{-\frac{y}{10}} = \dfrac{1}{60,000} y^3 e^{-\frac{y}{10}} & y > 0 \\ 0 & y \le 0 \end{cases}$$

4.115 $P(X < 4 | X \ge 2) = \dfrac{P(2 \le X < 4)}{P(X \ge 2)} = \dfrac{F(4) - F(2)}{1 - F(2)} = \dfrac{1 - e^{-4^2/3} - (1 - e^{-2^2/3})}{1 - (1 - e^{-2^2/3})}$

$$= \frac{e^{-4/3} - e^{-16/3}}{e^{-4/3}} = 1 - e^{-4} = 0.9817$$

4.117 Let c = budgeted amount in hundreds of dollars

$$0.10 = P(Y > c) = \int_c^1 3(1-y)^2 dy = -(1-y)^3 \Big|_c^1 = (1-c)^3$$

i.e., $c = 1 - \left(\dfrac{1}{10}\right)^{1/3} = 0.5358$; i.e., the budgeted amounts should be \$53.58.

4.119 For Y a gamma random variable with parameters $\beta = 1$, α, and X a Poisson random variable with parameter λ, the given relationship may be written as $P(Y > \lambda) = P(X \le \alpha - 1)$. Thus, for $\alpha = 2, \beta = 1, \lambda = 1$,

$$P(Y > 1) = P(X \le 1) = 0.736.$$

4.121 **a.** Let X = number of plants in a circular region of radius r. Then X has a Poisson distribution with parameters $\lambda\pi r^2$. Thus

$$P(R > r) = P(\text{no plants in an area of } \pi r^2) = P(X = 0)$$

$$= \frac{(\lambda\pi r^2)^0 e^{-\lambda\pi r^2}}{0!} = e^{-\lambda\pi r^2}.$$

Therefore, $F(r) = 1 - P(R > r) = 1 - e^{-\lambda\pi r^2}$ so that R has a Weibull distribution with parameters $\gamma = 2, \theta = \dfrac{1}{\lambda\pi}$.

b. $E(R) = \theta^{1/\gamma}\Gamma\left(1 + \dfrac{1}{\gamma}\right) = (\lambda\pi)^{-\frac{1}{2}}\Gamma\left(1 + \dfrac{1}{2}\right) = (\lambda\pi)^{-1/2}\dfrac{1}{2}\pi^{1/2} = \dfrac{1}{2\sqrt{\lambda}}$

4.123 Let X = gram weight of the certain type of aluminum. Then X has a lognormal distribution with parameters $\mu = 3, \sigma = 4$.

a. $E(X) = e^{\mu + \frac{\sigma^2}{2}} = e^{3 + \frac{4^2}{2}} = e^{11} = 59,874.142$ in units of 10^{-2} grams.

$$V(X) = e^{2\mu + \sigma^2}(e^{\sigma^2} - 1) = e^{2(3) + 4^2}(e^{4^2} - 1) = e^{38} - e^{22} \text{ in units of } 10^{-4}\text{g}^2.$$

b. By Tchebysheff's theorem, with $k = 2$, the required interval, in grams, is

$$[E(X) - 2\sqrt{V(X)}\ , E(X)2\sqrt{V(X)}]$$

$$= [e^{11} \times 10^{-2} - 2\sqrt{(e^{38} - e^{22}) \times 10^{-4}}\ , e^{11} \times 10^{-2} + 2\sqrt{(e^{38} - e^{22}) \times 10^{-4}}]$$

$$= (-3,569,047.08,\ \ 3,570,244.56)$$

Since gram weights are nonnegative, the interval becomes (0, 3, 570,244.56).

c. $P(X < 3) = P(Y < \ln 3) = P\left(Z < \frac{\ln 3 - 3}{4}\right) \approx P(Z < -0.48) = 0.5 - 0.1844 = 0.3156$

4.125 **a.** $X = \min(Y, T)$, where Y has an exponential distribution with mean 100, and $T = 200$. Hence

$$F_X(x) = \begin{cases} 0 & x \le 0 \\ 1 - e^{-x/100} & 0 < x < 200 \\ 1 & x \ge 200 \end{cases}$$

b. $E(X) = \int_0^\infty [1 - F_X(t)]\,dt = \int_0^{200} [1 - (1 - e^{-t/100})]\,dt$

$$= -100e^{-t/100}\Big|_0^{200} = 100(1 - e^{-2}) = 86.4665$$

4.127 $\Gamma(u) = \int_0^\infty \underbrace{y^{u-1}}_{w}\ \underbrace{e^{-y}dy}_{dv} = \underbrace{y^{u-1}(-e^{-y})\Big|_0^\infty}_{wv} - \underbrace{\int_0^\infty -e^{-y}(u-1)y^{u-2}dy}_{v\,dw}$

$$= (u-1)\int_0^\infty y^{u-2}e^{-y}dy = (u-1)\Gamma(u-1)$$

4.129 **a.** $B(\alpha, \beta) = \int_0^1 y^{\alpha-1}(1-y)^{\beta-1}dy = \int_0^{\pi/2} (\sin^2\theta)^{\alpha-1}(1 - \sin^2\theta)^{\beta-1} 2\sin\theta\cos\theta\, d\theta$

$$= 2\int_0^{\pi/2} (\sin\theta)^{2\alpha-1}(\cos\theta)^{2\beta-1}d\theta$$

since $1 - \sin^2\theta = \cos^2\theta$.

b. $\Gamma(\alpha)\Gamma(\beta) = \int_0^\infty y^{\alpha-1}e^{-y}dy\int_0^\infty x^{\beta-1}e^{-x}dx$

$$= \int_0^\infty (v^2)^{\alpha-1}e^{-v^2}2vdv\int_0^\infty (w^2)^{\beta-1}e^{-w^2}2wdw$$

$$= \int_0^\infty\int_0^\infty 4vw(v^2)^{\alpha-1}(w^2)^{\beta-1}e^{-(v^2+w^2)}dvdw$$

$$= \int_0^{\pi/2}\int_0^\infty 4r^2\sin\theta\cos\theta(r\sin\theta)^{2\alpha-2}(r\cos\theta)^{2\beta-2}e^{-r^2}rdrd\theta$$

$$= 2\int_0^{\pi/2}(\sin\theta)^{2\alpha-1}(\cos\theta)^{2\beta-1}\int_0^\infty 2r^{2(\alpha+\beta-1)}e^{-r^2}rdrd\theta$$

$$= 2\int_0^{\pi/2}(\sin\theta)^{2\alpha-1}(\cos\theta)^{2\beta-1}\int_0^\infty k^{\alpha+\beta-1}e^{-k}dkd\theta$$

$$= 2\int_0^{\pi/2}(\sin\theta)^{2\alpha-1}(\cos\theta)^{2\beta-1}\Gamma(\alpha+\beta)\,d\theta = \Gamma(\alpha+\beta)B(\alpha,\beta)$$

i.e.,

$$B(\alpha,\beta) = \frac{\Gamma(\alpha)\Gamma(\beta)}{\Gamma(\alpha+\beta)}$$

CHAPTER 5
MULTIVARIATE PROBABILITY DISTRIBUTIONS

5.1 Firms

Contract 1	Contract 2	Probability	(x_1, x_2)
I	I	$\frac{1}{9}$	(2,0)
I	II	$\frac{1}{9}$	(1,1)
I	III	$\frac{1}{9}$	(1,0)
II	I	$\frac{1}{9}$	(1,1)
II	II	$\frac{1}{9}$	(0,2)
II	III	$\frac{1}{9}$	(0,1)
III	I	$\frac{1}{9}$	(1,0)
III	II	$\frac{1}{9}$	(0,1)
III	III	$\frac{1}{9}$	(0,0)
(a,b)			

		x_1		
		0	1	2
	0	$\frac{1}{9}$	$\frac{2}{9}$	$\frac{1}{9}$
x_2	1	$\frac{2}{9}$	$\frac{2}{9}$	0
	2	$\frac{1}{9}$	0	0
$p(x_1)$		$\frac{4}{9}$	$\frac{4}{9}$	$\frac{1}{9}$

c. $P(X_1 = 1 \mid X_2 = 1) = \dfrac{P(X_1\ 1, X_2\ 1)}{P(X_2\ 1)} = \dfrac{2/9}{2/9 + 2/9} = \dfrac{1}{2}$

5.3 a.

		X_1	
		0	1
	0	0.0635	0.0775
	1	0.1007	0.0556
X_2	2	0.1630	0.0653
	3	0.1691	0.0549
	4	0.1929	0.0574

b.

X_1	$P(X_1 \mid X_2 = 0)$	$P(X_1 \mid X_2 = 1)$	$P(X_1 \mid X_2 = 2)$
0	0.4502	0.6455	0.7139
1	0.5498	0.3558	0.2861

X_1	$P(X_1 \mid X_2 = 3)$	$P(X_1 \mid X_2 = 4)$
0	0.7548	0.7707
1	0.2452	0.2293

Obviously, the older the child is, the better his/her chance of survival in a car accident without wearing a seat belt.

c.

X_2	$P(X_2 \mid X_1 = 0)$	$P(X_2 \mid X_1 = 1)$
0	0.0921	0.2495
1	0.1461	0.1788
2	0.2365	0.2102
3	0.2453	0.1768
4	0.2799	0.1847

No, this implies that if a child survives, then s/he will probably be older.

5.5 **a.** $f_1(x_1) = \begin{cases} \int_0^1 1\,dx_2 = 1 & 0 \le x_1 \le 1 \\ 0 & \text{otherwise} \end{cases}$

b. $P(X_1 \le 1/2) = \int_0^{1/3} 1\,dx_1 = 1/2$

c. $f_1(x_1) f_2(x_2) = 1 = f(x_1, x_2)$, for $0 \le x_1 \le 1, 0 \le x_2 \le 1$. Therefore, X_1 and X_2 are independent.

5.7 **a.** $P\left(X_1 \le \frac{3}{4}, X_2 \le \frac{3}{4}\right) = \int_0^{1/4}\int_0^{3/4} 2\,dx_2 dx_1 + \int_{1/4}^{3/4}\int_0^{1-x_1} 2\,dx_2 dx_1$

$$= 2\left(\frac{1}{4}\right)\left(\frac{3}{4}\right) + \int_{1/4}^{3/4} 2(1-x_1)\,dx_1 = \frac{3}{8} + \frac{1}{2} = \frac{7}{8}$$

Probability region of interest:

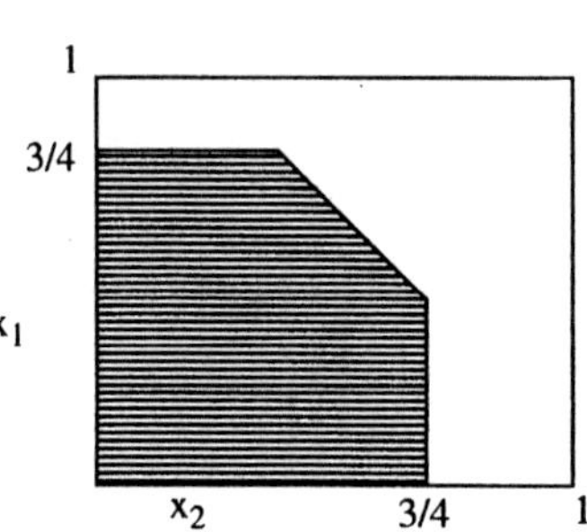

b. $P\left(X_1 \le \frac{1}{2}, X_2 \le \frac{1}{2}\right) = 2\left(\frac{1}{2}\right)\left(\frac{1}{2}\right) = \frac{1}{2}$

c. $P\left(X_1 \le \frac{1}{2} \middle| X_2 \le \frac{1}{2}\right) = \dfrac{P\left(X_1 \le \frac{1}{2}, X_2 \le \frac{1}{2}\right)}{P\left(X_2 \le \frac{1}{2}\right)}$

$$= \frac{1/2}{\int_0^{1/2}\int_0^{1-x_2} 2\,dx_1 dx_2} = \frac{1}{2\int_0^{1/2} 2(1-x_2)\,dx_2} = \frac{2}{3}$$

5.9 **a.** $P\left(X_1 < \frac{1}{2}, X_2 > \frac{1}{4}\right) = \int_0^{1/2}\int_{1/4}^{1} (x_1 + x_2)\, dx_2 dx_1 = \int_0^{1/2}\left(\frac{3x_1}{4} + \frac{15}{32}\right)dx_1 = \frac{21}{64}$

b. $P(X_1 + X_2 \le 1) = \int_0^1\int_0^{1-x_1} (x_1 + x_2)\, dx_2 dx_1 = \int_0^1 \left(x_1x_2 + \frac{x_2^2}{2}\right)\Bigg|_0^{1-x_1} dx_1$

$$= \int_0^1 \frac{1}{2}(1 - x_1^2)\, dx_1 = \frac{1}{3}$$

c. Note that

$$f_1(x_1)f_2(x_2) = \left(x_1 + \frac{1}{2}\right)\left(x_2 + \frac{1}{2}\right) \ne x_1 + x_2 = f(x_1, x_2)$$

for $0 \le x_1 \le 1, 0 \le x_2 \le 1$.

Therefore, X_1 and X_2 are not independent.

5.11 Note that

$$f_1(x_1) = \int_0^\infty \frac{1}{8}x_1 e^{-\frac{x_1}{2}} e^{-\frac{x_2}{2}} dx_2 = \frac{1}{4}x_1 e^{-\frac{x_1}{2}}\left(-e^{-\frac{x_2}{2}}\right)\Bigg|_0^\infty = \frac{1}{4}x_1 e^{-\frac{x_1}{2}} \text{ for } 0 < x_1 < \infty$$

and, integrating by parts, we have

$$f_2(x_2) = \int_0^\infty \frac{1}{8}x_1 e^{-\frac{x_1}{2}} e^{-\frac{x_2}{2}} dx_1 = \frac{1}{8}e^{-\frac{x_2}{2}}\left\{-2x_1 e^{-\frac{x_1}{2}}\Bigg|_0^\infty + 2\int_0^\infty e^{-\frac{x_1}{2}} dx_1\right\} = \frac{1}{8}e^{-\frac{x_2}{2}}\left\{-4e^{-\frac{x_1}{2}}\Bigg|_0^\infty\right\}$$

$$= \frac{1}{2}e^{-\frac{x_2}{2}} \text{ for } 0 < x_2 < \infty. \; x_1$$

a. Note that

$$f_1(x_1)f_2(x_2) = \frac{1}{4}x_1 e^{-\frac{x_1}{2}} \frac{1}{2}e^{-\frac{x_2}{28}} = \frac{1}{8}x_1 e^{-\frac{(x_1 + x_2)}{2}} = f(x_1, x_2)$$

for $0 < x_1 < \infty, 0 < x_2 < \infty$

Therefore, X_1 and X_2 are independent.

b. $P(X_1 > 1, X_2 > 1) = P(X_1 > 1)P(X_2 > 1) = \int_1^\infty \frac{1}{4}x_1 e^{-\frac{x_1}{2}} dx_1 \int_1^\infty \frac{1}{2}e^{-\frac{x_2}{2}} dx_2$

$$= \left\{-\frac{1}{2}x_1 e^{-\frac{x_1}{2}}\Bigg|_1^\infty + \frac{1}{2}\int_1^\infty e^{-\frac{x_1}{2}} dx_1\right\}\left\{-e^{-\frac{x_2}{2}}\Bigg|_1^\infty\right\}$$

$$= \left\{\frac{1}{2}e^{-\frac{1}{2}} + e^{-\frac{1}{2}}\right\}e^{-\frac{1}{2}} = \frac{3e^{-1}}{2} = 0.5518$$

5.13 Let X_i = arrival time for friend i, $0 \le x_i \le 1$, $i = 1, 2$. The two friends will meet if $|x_1 - x_2| < \frac{1}{3}$, and

$$P\left(|X_1 - X_2| < \frac{1}{3}\right) = \int_0^{1/6}\int_0^{x_1+1/6} 1\,dx_2dx_1 + \int_{1/6}^{5/6}\int_{x_1-1/6}^{x_1+1/6} 1\,dx_2dx_1$$

$$= \frac{3}{72} + \frac{2}{9} + \frac{3}{72} = \frac{11}{36}.$$

Probability region of interest:

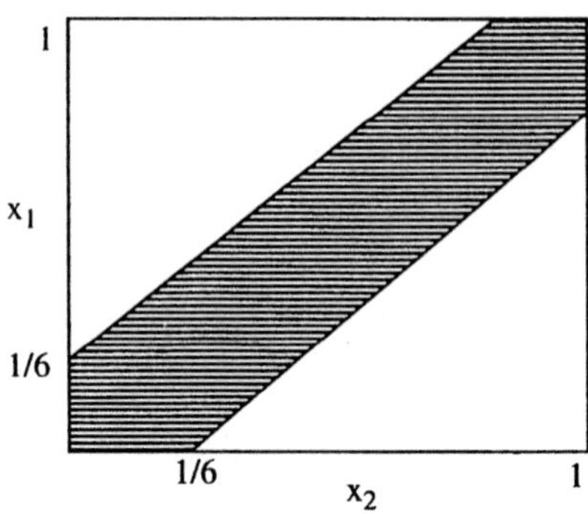

5.15 Let X_i = the time the ith call is made, $i = 1, 2$, $0 \le x_i \le 1$. Then

$$f(x_1, x_2) = f_1(x_1)f_2(x_2) = \begin{cases} 1 & 0 \le x_1 \le 1, 0 \le x_2 \le 1 \\ 0 & \text{otherwise} \end{cases}$$

a. $P\left(X_1 < \frac{1}{2}, X_2 < \frac{1}{2}\right) = P\left(X < \frac{1}{2}\right)P\left(X_2 < \frac{1}{2}\right) = \left(\frac{1}{2}\right)\left(\frac{1}{2}\right) = \frac{1}{4}$

b. $P\left(|X_1 - X_2| < \frac{1}{12}\right) = \int_0^{1/12}\int_0^{x_1+\frac{1}{12}} dx_2dx_1 + \int_{1/12}^{11/12}\int_{x_1-\frac{1}{12}}^{x_1+\frac{1}{12}} dx_2dx_1 + \int_{11/12}^{1}\int_{x_1-\frac{1}{12}}^{1} dx_2dx_1$

$$= \frac{3}{288} + \frac{20}{144} + \frac{3}{288} = \frac{23}{144}$$

Probability region of interest:

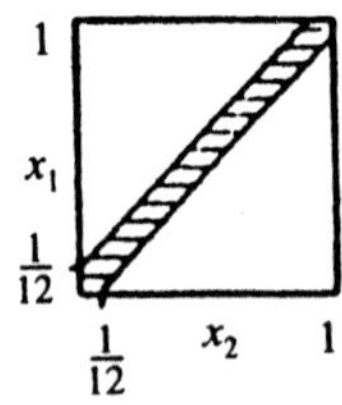

5.17 $f_{X_1}(x) = \{ \begin{matrix} 0.76 \text{ if } x & 0 \\ 0.24 \text{ if } x & 1 \end{matrix}$ and $f_{X_2}(x) = \begin{cases} 0.55 \text{ if } x & 0 \\ 0.16 \text{ if } x & 1 \\ 0.29 \text{ if } x & 2 \end{cases}$

a. $E(X_1) = 0(0.76)\,1(0.24) = 0.24$

$\text{Var}(X_1) = E(X_1^2) - (EX_1)^2$

$E(X_1^2) = 0^2(0.76) + 1^2(0.24) = 0.24$

So, $\text{Var}(X_1) = 0.24 - (0.24)^2 = 0.1824$

$E(X_2) = 0(0.55) + 1(0.16) + 2(0.29) = 0.74$

$\text{Var} = E(X_2^2) - (EX_2)^2$

$E(X_2^2) = 0^2(0.55) + 1^2(0.16) + 2^2(0.29) = 1.32$

So, $\text{Var}(X_2) = 1.32 - (0.74)^2 = 0.7724$

b. $\text{Cov}(X_1, X_2) = E(X_1X_2) - EX_1EX_2$

$E(X_1X_2) = (0)(0)(0.38) + (0)(1)(0.17) + (1)(0)(0.14) + (1)(1)(0.02)$

$+ (2)(0)(0.24) + (2)(1)(0.05) = 0.12$[230z

So, $\text{Cov}(X_1, X_2) = (0.12) - (0.24)(0.74) = -0.0576$

c. $E(Y) = E(X_1 + X_2) = 0.98$

$V(Y) = 0.1824 + 0.7724 + 0.1152 = 1.07$

5.19 **a.** $E(X_1 + X_2) = \int_0^1 \int_0^{1-x_1} (x_1 + x_2)\, 2dx_2dx_1 = \int_0^1 = (1 - x_1^2)\, dx_1 = \frac{2}{3}$

$E[(X_1 + X_2)^2] = \int_0^1 \int_0^{1-x_1} (x_1 + x_2)^2 2dx_2dx_1 = \int_0^1 \frac{2}{3}(1 - x_1^3)\, dx_1$

$= \frac{2}{3}\left(x - \frac{x^4}{4}\right)\Bigg|_1^0 = \frac{2}{3}\left(\frac{3}{4}\right) = \frac{1}{2}$

$V(Y) = E(Y^2) - [E(Y)]^2 = \frac{1}{2} - \left(\frac{2}{3}\right)^2 = \frac{1}{18}$

b. Using Tchebysheff's theorem with $k = \sqrt{2}$, we have $P(2/3 - \sqrt{2/18} < X_1 + X_2 < 2/3 + \sqrt{2/18}) \geq 0.5$; i.e., the desired interval is (1/3, 1).

5.21 **a.** $P(Y_1 < 2, Y_2 > 1) = \int_1^2 \int_{y_2}^2 e^{-y_1} dy_1 dy_2 = \int_1^2 (e^{-y_2} - e^{-2}) dy_2 = e^{-1} - 2e^{-2} = 0.0972$

Probability region of interest:

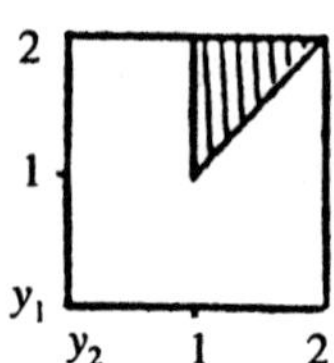

b. $P(Y_1 > 2Y_2) = \int_0^\infty \int_{2y_2}^\infty e^{-y_1} dy_1 dy_2 = \int_0^\infty e^{-2y_2} dy_2 = \frac{1}{2}$

c. $P(Y_1 - Y_2 \geq 1) = \int_0^\infty \int_{1+y_2}^\infty e^{-y_1} dy_1 dy_2 = \int_0^\infty e^{-(1+y_2)} dy_2 = e^{-1}$

d. $f_1(y_1) = \begin{cases} \int_0^{y_1} e^{-y_1} dy_2 = y_1 e^{-y_1} & 0 \leq y_1 < \infty \\ 0 & \text{otherwise} \end{cases}$

$f_2(y_2) = \begin{cases} \int_{y_2}^\infty e^{-y_1} dy_1 = y_1 e^{-y_2} & 0 \leq y_2 < \infty \\ 0 & \text{otherwise} \end{cases}$

5.23 **a.** $E(Y_1 - Y_2) = \int_0^\infty \int_{y_2}^\infty (y_1 - y_2) e^{-y_1} dy_1 dy_2 = \int_0^\infty e^{-y_2} dy_2 = 1$

b. $E[(Y_1 - Y_2)^2] = \int_0^\infty \int_{y_2}^\infty (y_1 - y_2)^2 e^{-y_1} dy_1 dy_2 = \int_0^\infty 2e^{-y_2} dy_2 = 2$

$V(Y) = E(Y^2) - [E(Y)]^2 = 2 - 1^2 = 1$

c. No, since

$$P(Y_1 - Y_2 > 2) = \int_0^\infty \int_{2+y_2}^\infty e^{-y_1} dy_1 dy_2 = \int_0^\infty e^{-(2+y_2)} dy_2 = e^{-2} = 0.1353$$

5.25 $\text{Cov}(X_1, X_2)$

$= E[(X_1 - \mu_1)(X_2 - \mu_2)]$

$= \int_{-\infty}^\infty \int_{-\infty}^\infty (x_1 - \mu_1)(x_2 - \mu_2) f(x_1, x_2)\, dx_1 dx_2$

$= \int_{-\infty}^\infty \int_{-\infty}^\infty (x_1 x_2 - x_1\mu_2 - x_2\mu_1 + \mu_1\mu_2) f(x_1, x_2)\, dx_1 dx_2$

$= \int_{-\infty}^\infty \int_{-\infty}^\infty (x_1 x_2) f(x_1, x_2)\, dx_1 dx_2 - \mu_2 \int_{-\infty}^\infty \int_{-\infty}^\infty x_1 f(x_1, x_2)\, dx_1 dx_2$

$-\mu_1 \int_{-\infty}^\infty \int_{-\infty}^\infty x_2 f(x_1, x_2)\, dx_1 dx_2 + \mu_1\mu_2 \int_{-\infty}^\infty \int_{-\infty}^\infty f(x_1, x_2)\, dx_1 dx_2$

$= E(X_1 X_2) - \mu_1\mu_2 - \mu_1\mu_2 + \mu_1\mu_2 = E(X_1 X_2) - \mu_1\mu_2$

5.27 Let $C = 20{,}000Y_1 + 10{,}000Y_2 + 2{,}000Y_3$, where Y_1, Y_2, Y_3 are as defined in Exercise 5.26. Then

$$\begin{aligned} E(C) &= 20{,}000E(Y_1) + 10{,}000E(Y_2) + 2{,}000E(Y_3) \\ &= 20{,}000np_1 + 10{,}000np_2 + 2{,}000np_3 \\ &= 20{,}000(4)(0.73) + 10{,}000(4)(0.20) + 2{,}000(4)(0.07) = 66{,}960 \end{aligned}$$

5.29 Let Y_1 = number of days having total radiation of at most 5 calories, Y_2 = number of days having total radiation between 5 and 6 calories, and Y_3 = number of days having radiation between 6 and 8 calories. Then (Y_1, Y_2, Y_3) has a multinomial distribution with parameters $n = 6$, $p_1 = 0.30$, $p_2 = 0.60 - 0.30$, $p_3 = 1 - 0.60 = 0.40$. Therefore,

$$P(Y_1 = 3, Y_2 = 1, Y_3 = 2) = \frac{6!}{3!\,1!\,2!}(0.30)^3(0.30)^1(0.40)^2 = 0.07776$$

This solution is based on the assumption that measurements on different days are independent.

5.31 Let Y_1 = number of customers passing through gate A, Y_2 = number of customers passing through gate B, and Y_3 = number of customers passing through gate C. Then (Y_1, Y_2, Y_3) has a multinomial distribution with parameters $n = 4$, $p_1 = p_2 = p_3 = 1/3$.

a. $$P(Y_1 = 2, Y_2 = 1, Y_3 = 1) = \frac{4!}{2!\,1!\,1!}\left(\frac{1}{3}\right)^2\left(\frac{1}{3}\right)^1\left(\frac{1}{3}\right)^1 = \frac{4}{27}$$

b. P(all four select same gate)

$$\begin{aligned} &= P(Y_1 = 4, Y_2 = 0, Y_3 = 0) + P(Y_1 = 0, Y_2 = 4, Y_3 = 0) \\ &\quad + P(Y_1 = 0, Y_2 = 0, Y_3 = 4) \\ &= (3)\frac{4!}{4!\,0!\,0!}\left(\frac{1}{3}\right)^4\left(\frac{1}{3}\right)^0\left(\frac{1}{3}\right)^0 = \frac{1}{27} \end{aligned}$$

c. P(all three gates are used)

$$\begin{aligned} &= P(Y_1 = 2, Y_2 = 1, Y_3 = 1) + P(Y_1 = 1, Y_2 = 2, Y_3 = 1) \\ &\quad + P(Y_1 = 1, Y_2 = 1, Y_3 = 2) \\ &= (3)\frac{4!}{1!\,2!\,1!}\left(\frac{1}{3}\right)^1\left(\frac{1}{3}\right)^2\left(\frac{1}{3}\right)^1 = \frac{4}{9} \end{aligned}$$

5.33 Let Y_0 = number out of the selected items with no defects. Then (Y_0, Y_1, Y_2) has a multinomial distribution with parameters $n = 10$, $p_0 = 0.85$, $p_1 = 0.10$, $p_2 = 0.05$, and $E(Y_i) = np_i$, $V(Y_i) = np_i(1 - p_i)$, $\text{Cov}(Y_i, Y_j) = -np_ip_j$, $i,j = 1, 2, 3$, $i \neq j$. Therefore we have

$$E(Y_1 + 3Y_2) = E(Y_1) + 3E(Y_2) = 10(0.10) + 3(10)(0.05) = 2.5$$

and

$$V(Y_1 + 3Y_2) = V(Y_1) + (3)^2 V(Y_2) + 2(3)\text{Cov}(Y_1, Y_2)$$

$$= 10(0.10)(0.90) + 9(10)(0.05)(0.95) - 2(3)(10)(0.10)(0.05) = 4.875.$$

5.35 Let Y_1 = number of vehicles that turn left, Y_2 = number of vehicles that turn right, and Y_3 = number of vehicles that continue straight, out of n vehicles arriving at the intersection. Then (Y_1, Y_2, Y_3) has a multinomial distribution with parameters n, $p_1 = 0.4$, $p_2 = 0.25$, $p_3 = 0.35$.

a. $P(Y_1 = 1, Y_2 = 1, Y_3 = 3) = \dfrac{5!}{1!1!3!}(0.4)(0.25)(0.35)^3 = 0.08575$

b. Note that the marginal distribution of Y_2 is binomial with parameters $n = 5$, $p = 0.25$, so

$$P(Y_2 \geq 1) = 1 - P(Y_2 = 0) = 1 - \binom{5}{0}(0.25)^0(0.75)^5 = 1 - (0.75)^5 = 0.7627.$$

c. Note that, assuming vehicles turn independently of each other, the marginal distribution of Y_1 is binomial with parameters $n = 100$, $p = 0.4$, so

$$E(Y_1) = np = 100(0.4) = 40$$

$$V(Y_1) = np(1 - p) = 100(0.4)(0.6) = 24.$$

5.37 Let X_1 = number of cars arriving at entrance I, X_2 = number of cars arriving at entrance II, and $Y = X_1 + X_2$. Then X_1 has a Poisson distribution with parameter $\lambda_1 = 3$, X_2 has a Poisson distribution with parameter $\lambda_2 = 4$, and Y has a Poisson distribution with parameter $\lambda_Y = \lambda_1 + \lambda_2 = 3 + 4 = 7$, and

$$P(Y = 3) = \frac{\lambda_Y^3 e^{-\lambda_Y}}{3!} = \frac{7^3 e^{-7}}{6} = 0.05213.$$

5.39 Let X = resistance of resistor. Then X has a normal distribution with parameters $\mu_X = 100$, $\sigma_X^2 = 100$, and, from Exercise 5.38, $Y = 2X$ has a normal distribution with parameters $\mu_Y = 2\mu_X = 2(100) = 200$, and $\sigma_Y^2 = 2^2\sigma_X^2 = 4(100) = 400$.

a. $P(Y > 220) = P\left(\dfrac{Y - 200}{\sqrt{400}} > \dfrac{220 - 200}{\sqrt{400}}\right) = P(Z > 1) = 0.5 - 0.3413 = 0.1587$

b. $P(Y < 190) = P\left(\dfrac{Y - 200}{\sqrt{400}} < \dfrac{190 - 200}{\sqrt{400}}\right) = P\left(Z < -\dfrac{1}{2}\right) = 0.5 - 0.1915 = 0.3085$

5.41 $Y \sim P(\lambda)$ and $\lambda \sim \Gamma(\alpha, \beta)$. From Theorem 5.3,

$$E(Y) = E[E(Y|\lambda)]$$
$$= E[\lambda] \quad \text{(from section 3.7)}$$
$$= \alpha\beta \quad \text{(from section 4.5)}$$

And from Theorem 5.4

$$\text{Var}(Y) = E[\text{Var}(Y|\lambda)] + \text{Var}[E(Y|\lambda)]$$
$$= E[\lambda] + \text{Var}[\lambda] \quad \text{(from section 3.7)}$$
$$= \alpha\beta + \alpha\beta^2 \quad \text{(from section 4.5)}$$
$$= \alpha\beta(1+\beta)$$

5.43 **a.** $E(Y^2|X = 1)$ = "Expected number of trials until first success squared given that the first trial was a success"

= $1^2 = 1$

$E(Y^2|X = 0)$ = "Expected number of trials until first success squared given that the first trial was not a success"

= "Expected number of trials until first success plus one trial squared"

= $E(Y+1)^2$

b. $E(Y^2) = E(E(Y^2|X)) = \sum_{x=0}^{1} E(Y^2|X = x)P[X - x]$

c. $E(Y^2) = E(E(Y^2|X)) = \sum_{x=0}^{1} E(Y^2|X = x)P[X = x]$

$$= E(Y^2|X = 0)P[X = 0] + E(Y^2|X = 1)P[X = 1]$$
$$= E(Y+1)^2(1-p) + (1)(p)$$
$$= E(Y^2+2Y+1)(1-p) + p$$
$$= (1-p)EY^2 + 2(1-p)EY + 1 - p + p$$
$$= EY^2 - pEY^2 + 2(1-p)1/p + 1$$
$$= EY^2 - pEY^2 + 2/p - 2 + 1$$
$$= EY^2 - pEY^2 + 2/p - 1.$$

So,

$$EY^2 = EY^2 - pEY^2 + 2/p - 1$$

which implies that

$$pEY^2 = 2/p - 1$$

or that

$$p^2EY^2 = 2 - p.$$

Hence,

$$EY^2 = \frac{2-p}{p^2}$$

d. $V(Y) = E(Y^2) - (EY)^2$

$$= \frac{2-p}{p^2} - \left(\frac{1}{p}\right)^2$$

$$= \frac{2-p-1}{p^2} = \frac{1-p}{p^2}$$

5.45 **a.** $f_1(x_1) = \int_{\infty}^{x_1} f(x_1, x_2)\, dx_2 = \begin{cases} \int_0^{x_1} 3x_1 dx_2 = 3x_1^2 & 0 \le x_1 \le 1 \\ 0 & \text{otherwise} \end{cases}$

$$f_2(x_2) = \begin{cases} \int_{x_2}^{1} 3x_1 dx_1 = \frac{3}{2}(1 - x_2^2) & 0 \le x_2 \le 1 \\ 0 & \text{otherwise} \end{cases}$$

b. $P\left(X_1 \le \frac{3}{4}, X_2 \le \frac{1}{2}\right) = \int_0^{1/2}\int_{x_2}^{3/4} 3x_1 dx_1 dx_2 = \int_0^{1/2} \frac{3}{2}\left(\frac{9}{16} - x_2^2\right) dx_2$

$$= \frac{3}{2}\left[\frac{9}{32} - \frac{1}{3}\left(\frac{1}{2}\right)^3\right] = \frac{23}{64}$$

c. $P\left(X_1 \le \frac{1}{2} \middle| X_2 \ge \frac{3}{4}\right) = \dfrac{P\left(X_1 \le \frac{1}{2}, X_2 \ge \frac{3}{4}\right)}{P\left(X_2 \ge \frac{3}{4}\right)} = \dfrac{0}{P\left(X_2 \ge \frac{3}{4}\right)} = 0$

5.47 $f(x_1 | X_2 = x_2) = \dfrac{f(x_1, x_2)}{f(x_2)} = \dfrac{4x_1x_2}{2x_2} = 2x_1$ for $0 \le x_1 \le 1$. Then X_1 and X_2 are independent, since $f(x_1 | X_2 = x_2) = f_1(x_1) = 2x_1$ for $0 \le x_1 \le 1$.

$0 \le y \le x \le 1$
$X \sim \text{Unif}$

5.49 **a.** $f(x_2|x_1) = \begin{cases} \frac{1}{x_1} & 0 \le x_2 \le x_1 \le 1 \\ 0 & \text{otherwise} \end{cases}$

$f(x_1) = \begin{cases} 1 & 0 \le x_1 \le 1 \\ 0 & \text{otherwise} \end{cases}$

Therefore,

$f(x_1, x_2) = f(x_2|x_1)f(x_1)$

$\begin{cases} \frac{1}{x_1} & 0 \le x_2 \le x_1 \le 1 \\ 0 & \text{otherwise} \end{cases}$

b. $P(x_2 \ge \frac{1}{4} \Big| X_1 = \frac{1}{2}) = \int_{1/4}^{\infty} f(x_2|x_1 = \frac{1}{2})\,dx_2 = \int_{1/4}^{1/2} \frac{1}{1/2}\,dx_2 = \frac{1}{2}$

c. $f(x_1|x_2) = \frac{f(x_1, x_2)}{f(x_2)}$

$$= \begin{cases} \dfrac{\frac{1}{x_1}}{\int_{x_2}^{1} \frac{1}{x_1}\,dx_1} = \dfrac{-1}{x_1 \ln x_2} & 0 \le x_2 \le x_1 \le 1 \\ 0 & \text{otherwise} \end{cases}$$

So $P\left(X_1 \ge \frac{1}{2} \Big| X_2 = \frac{1}{4}\right) = \int_{1/2}^{1} -\frac{1}{x_1 \ln\left(\frac{1}{4}\right)}\,dx_1 = \frac{1}{\ln 4}\left(\ln x_1 \Big|_{1/2}^{1}\right) = \frac{\ln 2}{\ln 4}$

5.51 **a.** $f(x_1) = \begin{cases} \int_0^1 (x_1 + x_2)\,dx_2 = x_1 + \frac{1}{2} & 0 \le x_1 \le 1 \\ 0 & \text{otherwise} \end{cases}$

$f(x_2) = \begin{cases} \int_0^1 (x_1 + x_2)\,dx_1 = x_2 + \frac{1}{2} & 0 \le x_2 \le 1 \\ 0 & \text{otherwise} \end{cases}$

b. $f(x_1)f(x_2) = \left(x_1 + \frac{1}{2}\right)\left(x_2 + \frac{1}{2}\right) \ne f(x_1, x_2) = x_1 + x_2$ for $0 \le x_1, x_2 \le 1$. Therefore, X_1 and X_2 are not independent.

c. $f(x_1|x_2) = \frac{f(x_1, x_2)}{f(x_2)}$

$$= \begin{cases} \dfrac{x_1 + x_2}{x_2 + \frac{1}{2}} & 0 \le x_1, x_2 \le 1 \\ 0 & \text{otherwise} \end{cases}$$

5.53 Note that $f(x_1, x_2) = 1$, for $0 \le x_2 + |x_1| \le 1, 0 \le x_2 \le 1, |x_1| \le 1$.

a. $$f(x_2) = \begin{cases} \int_{x_2-1}^{1-x_2} 1\,dx_1 = 2(1-x_2) & 0 \le x_2 \le 1 \\ 0 & \text{otherwise} \end{cases}$$

b. $$f(x_1) = \begin{cases} \int_{0}^{1-|x_1|} 1\,dx_2 = 1-|x_1| & |x_1| \le 1 \\ 0 & \text{otherwise} \end{cases}$$

c. $$P[(X_1 - X_2) \ge 0] = \int_0^{1/2}\int_{x_2}^{1-x_2} 1\,dx_1dx_2 = \int_0^{1/2} (1-2x_2)\,dx_2$$

$$= \frac{1}{2} - \left(\frac{1}{2}\right)^2 = \frac{1}{4}$$

Probability region of interest:

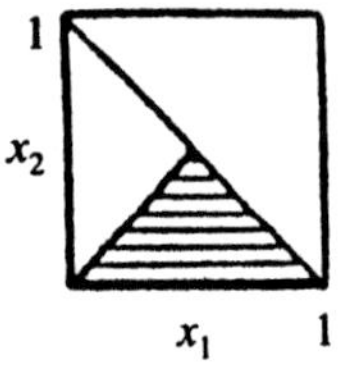

5.55 $$E(X_1) = \int_0^1 x_1 3x_1^2\,dx_1 = \frac{3}{4}$$

$$E(X_2) = \int_0^1 x_2\left(\frac{3}{2}\right)(1-x_2^2)\,dx_2 = \frac{3}{2}\left(\frac{1}{2} - \frac{1}{4}\right) = \frac{3}{8}$$

$$E(X_1X_2) = \int_0^1\int_0^{x_1} x_1x_2 3x_1\,dx_2dx_1 = \int_0^1 \frac{3x_1^4}{2}dx_1 = \frac{3}{10}$$

$$\text{Cov}(X_1, X_2) = E(X_1X_2) - E(X_1)E(X_2) = \frac{3}{10} - \frac{3}{4}\left(\frac{3}{8}\right) = \frac{3}{160}$$

5.57 **a.** $E(X_1X_2) = \int_0^1\int_0^1 x_1x_2(x_1+x_2)\,dx_1dx_2 = \int_0^1 x_2\left(\frac{1}{3}+\frac{x_2}{2}\right)dx_2 = \frac{1}{6}+\frac{1}{6} = \frac{1}{3}$

$$E(X_i) = \int_0^1 x_i\left(x_i+\frac{1}{2}\right)x_i = \frac{1}{3}+\frac{1}{4} = \frac{7}{12},\ \text{for } i = 1, 2$$

$$\text{Cov}(X_1, X_2) = E(X_1X_2) - E(X_1)E(X_2) = \frac{1}{3}-\frac{7}{12}\left(\frac{7}{12}\right) = \frac{-1}{144}$$

b. $E(3X_1-2X_2) = 3E(X_1) - 2E(X_2) = 3\left(\frac{7}{12}\right)-2\left(\frac{7}{12}\right) = \frac{7}{12}$

c. Note that

$$E(X_i^2) = \int_0^1 x_i^2\left(x_i+\frac{1}{2}\right)dx_i = \frac{1}{4}+\frac{1}{6} = \frac{5}{12},\ i = 1, 2$$

$$V(X_i^2) = \frac{5}{12}-\left(\frac{7}{12}\right)^2 = \frac{11}{144},\ i = 1, 2$$

so we have

$$V(3X_1-2X_2) = 9V(X_1)+4V(X_2)+2(3)(-2)\text{Cov}(X_1, X_2)$$

$$= 9\left(\frac{11}{144}\right)+4\left(\frac{11}{144}\right)-12\left(-\frac{1}{144}\right) = \frac{155}{144} = 1.0764.$$

5.59 Let X = number of defectives selected. Then $X|p$ has a binomial distribution with parameters $n = 3$, p, and

$$P(X = x|p) = \binom{n}{x}p^x(1-p)^{n-x}$$

$$P(X = x, p) = \binom{n}{x}p^x(1-p)^{n-x}$$

$$P(X = x) = \int_0^1\binom{n}{x}p^x(1-p)^{n-x}dp.$$

Therefore,

$$P(X=2) = \int_0^1\binom{3}{2}p^2(1-p)^{3-2}dp = \int_0^1 3p^2(1-p)\,dp = \int_0^1 3(p^2-p^3)\,dp$$

$$= 3\left(\frac{1}{3}-\frac{1}{4}\right) = \frac{1}{4}$$

5.61 Since X_1, X_2, and X_3 are independent, we have

$$f(x_1, x_2, x_3) = f(x_1)f(x_2)f(x_3)$$

$$= \begin{cases} \left(\frac{1}{\theta}\right)^3 e^{-\frac{1}{\theta}(x_1 + x_2 + x_3)} & \theta, x_1, x_2, x_3 > 0 \\ 0 & \text{otherwise} \end{cases}$$

5.63 Note that

$$E(X_i) = 0(1-p) + 1(p) = p$$

$$E(X_i^2) = 0(1-p) + 1(p) = p$$

$$V(X_i) = E(X_i^2) - [E(X_i)]^2 = p - p^2 = p(1-p)$$

and X_i is independent of $X_{i'}$, $i \neq i'$, $i = 1, 2, ..., n$, $i' = 1, 2, \ldots, n$. Then

$$E(Y) = E\sum_{i=0}^{n} X_i = \sum_{i=0}^{n} E(X_i) = \sum_{i=1}^{n} p = np$$

$$V(Y) = V\left(\sum_{i=0}^{n} X_i\right) = \sum_{i=0}^{n} V(X_i) = \sum_{i=1}^{n} p(1-p) = np(1-p)$$

5.65 **a.** $E(X|\lambda) = \lambda$. Then

$$E(X) = E[E(X|\lambda)]$$

$$= \int_{-\infty}^{\infty} E(X|\lambda)f(\lambda)\,d\lambda = \int_{0}^{\infty} \lambda e^{-\lambda} d\lambda = \Gamma(2) = 1$$

b. $$E(X) = \sum_{x=0}^{\infty} xf(x) = \sum_{x=0}^{\infty} x\left(\frac{1}{2}\right)^{x+1} = \frac{1}{2}\sum_{x=0}^{\infty} x\left(\frac{1}{2}\right)^{x}$$

$$= \frac{1}{2}\left\{\frac{\frac{1}{2}}{\left(\frac{1}{2}\right)^2}\right\} = 1$$

5.67 $F(x) = \int_0^x \frac{1}{200} e^{-\frac{y}{200}} dy = -e^{-\frac{y}{200}} \Big|_0^x = 1 - e^{-\frac{x}{200}}$

$$f(x|\, x > x_1) = \frac{f(x)}{1 - F(x_1)} = \frac{\frac{1}{200} e^{-\frac{x}{200}}}{1 - \left(1 - e^{\frac{-x_1}{200}}\right)} = \frac{1}{200} e^{-\frac{x - x_1}{200}}$$

$$E(X|\, X > x_1) = \int_{x_1}^{\infty} x f(x|\, x > x_1)\, dx = \int_{x_1}^{\infty} \frac{x}{200} e^{-\frac{x - x_1}{200}} dx$$

$$= e^{x_1/200} \left\{ \left(-x e^{-\frac{x}{200}} \Big|_{x_1}^{\infty} \right) + \int_{x_1}^{\infty} e^{-\frac{x}{200}} dx \right\}$$

$$= x_1 - e^{\frac{x_1}{200}} \left(200 e^{-\frac{x}{200}} \Big|_{x_1}^{\infty} \right) = 200 + x_1$$

Therefore,

$$E(X|\, X > 100) = 200 + 100 = 300.$$

5.69 **a.** $E(e^{t_1 X_1 + t_2 X_2 + t_3 X_3}) = \sum_{\substack{x_1, x_2, x_3 \\ x_1 + x_2 + x_3 = n}} e^{t_1 x_1 + t_2 x_2 + t_3 x_3} \frac{n!}{x_1! x_2! x_3!} p_1^{x_1} p_2^{x_2} p_3^{x_3}$

$$= \sum_{\substack{x_1, x_2, x_3 \\ x_1 + x_2 + x_3 = n}} \frac{n!}{x_1! x_2! x_3!} (p_1 e^{t_1})^{x_1} (p_2 e^{t_2})^{x_2} (p_3 e^{t_3})^{x_3}$$

$$= (p_1 e^{t_1} + p_2 e^{t_2} + p_3 e^{t_3})^n$$

b. $E(X_1 X_2) = \frac{\partial^2 M_{X_1, X_2, X_3}(t_1, t_2, 0)}{\partial t_1 \partial t_2} \Big|_{t_1 = t_2 = 0}$

$$= n p_1 (n-1) p_2 = n^2 p_1 p_2 - n p_1 p_2$$

$$E(X_1) = \frac{\partial M_{X_1, X_2, X_3}(t_1, 0, 0)}{\partial t_1} \Big|_{t_1 = 0}$$

$$= n p_1$$

Similarly,

$$E(X_2) = n p_2$$

$$\text{Cov}(X_1, X_2) = E(X_1 X_2) - E(X_1) E(X_2)$$

$$= n^2 p_1 p_2 - n p_1 p_2 - n p_1 n p_2 = -n p_1 p_2$$

5.71 $P(X_i = k) = \frac{1}{2}$ for $i = 1, 2, 3, k = 0, 1$, and

$$P(X_i = k, X_{i'} = k') = \frac{1}{4} = P(X_i = k)P(X_{i'} = k')$$

for $i \neq i'$, $i = 1, 2, 3$; $i' = 1, 2, 3$; $k = 0, 1$; $k' = 0, 1$. Therefore, X_i and $X_{i'}$ are independent, for $i \neq i'$. However,

$$P(X_1 = 1, X_2 = 1, X_3 = 1) = \frac{1}{4} \neq P(X_1 = 1)P(X_2 = 1)P(X_3 = 1) = \frac{1}{8}.$$

Therefore, X_1, X_2, X_3 are not jointly independent.

5.73 Note that

$$p(x_i) = \frac{\binom{N_i}{x_i}\binom{N-N_i}{n-x_i}}{\binom{N}{n}},$$

$i = 1, 2, 3$; i.e., X_i has a hypergeometric distribution, and $E(X_i) = np_i$, $V(X_i) = n\left(\frac{N_i}{N}\right)\left(\frac{N-N_i}{N}\right)\left(\frac{N-n}{N-1}\right)$. Also,

$$p(x_1, x_2) = \frac{\binom{N_1}{x_i}\binom{N_2}{x_2}\binom{N-N_1-N_2}{n-x_1-x_2}}{\binom{N}{n}}.$$

Then

$$E(X_1X_2) = \sum_{x_2=0}^{n}\sum_{x_1=0}^{n-x_2} x_1x_2\frac{\binom{N_1}{x_1}\binom{N_2}{x_2}\binom{N-N_1-N_2}{n-x_1-x_2}}{\binom{N}{n}}$$

$$= \sum_{x_2=0}^{n} x_2\frac{\binom{N_2}{x_2}}{\binom{N}{n}}\sum_{x_1=0}^{n-x_2} x_1\binom{N_1}{x_1}\binom{(N-N_2)-N_1}{(n-x_2)-x_1}$$

$$= N_1\sum_{x_2=0}^{n} x_2\frac{\binom{N_2}{x_2}}{\binom{N}{n}}\sum_{x_1=1}^{n-x_2}\binom{N_1-1}{x_1-1}\binom{(N-N_2)-N_1}{(n-x_2)-x_1}$$

$$= N_1 \sum_{x_2=0}^{n} x_2 \frac{\binom{N_2}{x_2}}{\binom{N}{n}} \sum_{y=0}^{n-x_2-1} \binom{N_1-1}{y} \binom{(N-N_2)-N_1}{(n-x_2)-1-y}$$

$$= N_1 \sum_{x_2=0}^{n} x_2 \frac{\binom{N_2}{x_2}}{\binom{N}{n}} \binom{N-N_2-1}{n-x_2-1} \left(\text{since} \sum_{i=0}^{m} \binom{a}{i}\binom{b}{m-i} = \binom{a+b}{m} \right)$$

$$= \sum_{x_2=1}^{n} N_1 N_2 \frac{\binom{N_2-1}{x_2-1}\binom{N-N_2-1}{n-x_2-1}}{\binom{N}{n}} = \sum_{x_2=1}^{n-1} N_1 N_2 \frac{\binom{N_2-1}{x_2-1}\binom{N-N_2-1}{n-x_2-1}}{\binom{N}{n}}$$

$$= N_1 N_2 \sum_{x=0}^{n-2} \frac{\binom{N_2-1}{z}\binom{N-N_2-1}{n-2-z}}{\binom{N}{n}} = N_1 N_2 \frac{\binom{N-2}{n-2}}{\binom{N}{n}}$$

$$= N_1 N_2 \frac{n(n-1)}{N(N-1)}$$

$$\text{Cov}(X_1 X_2) = E(X_1 X_2) - E(X_1)E(X_2) = N_1 N_2 \frac{n(n-1)}{N(N-1)} - n\left(\frac{N_1}{N}\right) n\left(\frac{N_2}{N}\right)$$

$$= \frac{nN_1N_2}{N}\left(\frac{n-1}{N-1} - \frac{n}{N}\right) = -\frac{nN_1N_2(N-n)}{N^2(N-1)}$$

$$\text{Corr}(X_1, X_2) = \frac{\text{Cov}(X_1, X_2)}{\sqrt{V(X_1)V(X_2)}}$$

$$= \frac{-\frac{nN_1N_2(N-n)}{N^2(N-1)}}{\sqrt{n\left(\frac{N_1}{N}\right)\left(\frac{N-N_1}{N}\right)\left(\frac{N-n}{N-1}\right) n\left(\frac{N_2}{N}\right)\left(\frac{N-N_2}{N}\right)\left(\frac{N-n}{N-1}\right)}} .$$

$$= -\sqrt{\frac{N_1 N_2}{(N-N_1)(N-N_2)}} = -\sqrt{\frac{p_1 p_2}{(1-p_1)(1-p_2)}}$$

5.75 **a.** $f(x_1, x_2) = f(x_1)f(x_2)$ (Since X_1 and X_2 are independent)

$$= \begin{cases} \left(\frac{1}{3}\right)e^{-\frac{x_1}{3}}\left(\frac{1}{3}\right)e^{-\frac{x_2}{3}} = \frac{e^{-(x_1+x_2)/3}}{9} & x_1 > 0, x_2 > 0 \\ 0 & \text{otherwise} \end{cases}$$

b. $$P(X_1 + X_2 \leq 1) = \int_0^1 \int_0^{1-x_1} \frac{e^{-(x_1+x_2)/3}}{9} dx_2 dx_1 = \int_0^1 \left(\frac{1}{3}\right)e^{-\frac{x_1}{3}}\left(-e^{-\frac{x_2}{3}}\Big|_0^{1-x_1}\right)dx_1$$

$y = 1 - x$

$$= \int_0^1 \left(\frac{1}{3}\right)e^{-\frac{x_1}{3}}\left(1 - e^{-\frac{1-x_1}{3}}\right)dx_1 = \frac{1}{3}\int_0^1 \left(e^{-\frac{x_1}{3}} - e^{-\frac{1}{3}}\right)dx_1$$

$$= \frac{1}{3}\left\{-3e^{-\frac{x_1}{3}}\Big|_0^1 - e^{-\frac{1}{3}}\right\} = 1 - \frac{4e^{-\frac{1}{3}}}{3} = 0.0446$$

$$f(x) = \begin{cases} (1/3)e^{-x/3} & \text{for } x>0 \\ 0 & \text{ow} \end{cases}$$

CHAPTER 6
FUNCTIONS OF RANDOM VARIABLES

6.1 **a.** $F_{U_1}(u) = P(U_1 \le u) = P(2X - 1 \le u) = P\left(X \le \frac{u+1}{2}\right)$

$$= F_X\left(\frac{u+1}{2}\right) = \int_0^{\frac{u+1}{2}} 2(1-x)\,dx = u + 1 - \frac{(u+1)^2}{4} = \frac{2u+3-u^2}{4}$$

for $0 \le \frac{u+1}{2} \le 1$; i.e., $-1 \le u \le 1$

$$f_{U_1}(u) = \frac{dF_{U_1}(u)}{du}$$

$$= \begin{cases} \dfrac{d\dfrac{2u+3-u^2}{4}}{du} = \dfrac{1}{2}(1-u) & -1 \le u \le 1 \\ 0 & \text{otherwise} \end{cases}$$

b. $F_{U_2} = P(U_2 \le u) = P(1 - 2X \le u) = P\left(X \ge \frac{1-u}{2}\right) = 1 - F_X\left(\frac{1-u}{2}\right)$

$$= 1 - \int_0^{(1-u)/2} 2(1-x)\,dx = 1 - \left[(1-u) - \frac{(1-u)^2}{4}\right] = \frac{u^2+2u+1}{4}$$

for $0 \le \frac{1-u}{2} \le 1$; i.e., $-1 \le u \le 1$

$$f_{U_2}(u) = \frac{dF_{U_2}(u)}{du}$$

$$= \begin{cases} \dfrac{d\dfrac{u^2+2u+1}{4}}{du} = \dfrac{1}{2}(u+1) & -1 \le u \le 1 \\ 0 & \text{otherwise} \end{cases}$$

c. $F_{U_3}(u) = P(U_3 \le u) = P(X^2 \le u) = P(X \le \sqrt{u})$

$$= F_X(\sqrt{u}) = \int_0^{\sqrt{u}} 2(1-x)\,dx = 2\sqrt{u} - u$$

for $0 \le \sqrt{u} \le 1$; i.e., $0 \le u \le 1$

$$f_{U_3}(u) = \frac{dF_{U_3}(u)}{du}$$

$$= \begin{cases} \dfrac{d(2\sqrt{u}-u)}{du} = \dfrac{1-\sqrt{u}}{\sqrt{u}} & 0 \le u \le 1 \\ 0 & \text{otherwise} \end{cases}$$

6.3 **a.** $F_U(u) = P(Y \le u) = P(10X - 4 \le u) = P\left(X \le \dfrac{u+4}{10}\right) = F_X\left(\dfrac{u+4}{10}\right)$

$$= \begin{cases} \displaystyle\int_0^{\frac{u+4}{10}} x\,dx = \frac{(u+4)^2}{200} & 0 \le \dfrac{u+4}{10} \le 1 \quad (\text{ie, } -4\text{a} \le u \le 6) \\ F_X(1) + \displaystyle\int_1^{\frac{u+4}{10}} dx = \frac{1}{2} + \frac{u+4}{10} - 1 = \frac{u-1}{10} & 1 < \dfrac{u+4}{10} \le 1.5 \quad (\text{i e } 6 < \text{u} \le 11) \end{cases}$$

$$f_U(u) = \frac{df_U(u)}{du}$$

$$= \begin{cases} \dfrac{d\dfrac{(u+4)^2}{200}}{du} = \dfrac{u+4}{100} & -4 \le u \le 6 \\ \dfrac{d\dfrac{u-1}{10}}{du} = \dfrac{1}{10} & 6 < u \le 11 \\ 0 & \text{otherwise} \end{cases}$$

b. $E(U) = \displaystyle\int_{-4}^{6} u\left(\frac{u+4}{100}\right)du + \int_6^{11} \frac{u}{10}du = \frac{1}{100}\left(\frac{u^3}{3} + 2u^2\right)\Bigg|_{-4}^{6} + \frac{u^2}{20}\Bigg|_6^{11}$

$$= \frac{1}{100}\left(\frac{6^3}{3} + 72 + \frac{64}{3} - 32\right) + \frac{1}{20}(11^2 - 36) = \frac{4}{3} + \frac{17}{4} = \frac{67}{12}$$

c. $E(X) = \displaystyle\int_0^1 x^2dx + \int_1^{1.5} xdx = \frac{1}{3} + \frac{1}{2}\left(\frac{9}{4} - 1\right) = \frac{1}{3} + \frac{9}{8} - \frac{1}{2} = \frac{23}{24}$

$$E(U) = E(10X - 4) = 10E(X) - 4 = \frac{230}{24} - 4 = \frac{67}{12}$$

6.5 **a.** $F_U(u) = P(U \le u) = P(X_1 - X_2 \le u) = P(X_1 \le u + X_2)$

for $0 \le u \le 1$:

$$= \int_0^u \int_{2x_2}^{x_2+u} dx_1 dx_2 = \int_0^u (x_2 + u - 2x_2)\, dx_2 = \int_0^u (u - x_2)\, dx_2 = u^2 - \frac{u^2}{2} = \frac{u^2}{2}$$

for $1 < u \le 2$:

$$= 1 - \int_0^{2-u} \int_{x_2+u}^{2} dx_1 dx_2 = 1 - \int_0^{2-u} (2 - x_2 - u)\, dx_2 = 1 - \frac{(2-u)^2}{2}$$

$$f_U(u) = \frac{dF_U(u)}{du}$$

$$= \begin{cases} \dfrac{d\frac{u^2}{2}}{du} = u & 0 \le u \le 1 \\ \dfrac{d\left(1 - \frac{(2-u)^2}{2}\right)}{du} = 2 - u & 1 < u \le 2 \\ 0 & \text{otherwise} \end{cases}$$

b. $E(U) = \int_0^1 u^2 du + \int_1^2 u(2-u)\, du = \frac{1}{3} + \left(u^2 - \frac{u^3}{3}\right)\Bigg|_1^2 = \frac{1}{3} + \left(4 - \frac{8}{3} - 1 + \frac{1}{3}\right) = 1$

This answer agrees with the result of Exercise 5.18.

6.7 **a.** $f(x_1, x_2) = 2$ for $0 \le x_1 \le 1, 0 \le x_2 \le 1, 0 \le x_1 + x_2 \le 1$

$$= F_U(u) = P(U \le u) = P(X_1 + X_2 \le u) = P(X_1 \le u - X_2)$$

$$= \int_0^u \int_0^{u-x_2} 2dx_1 dx_2 = 2\int_0^u (u - x_2)\, dx_2$$

$$= 2\left(u^2 - \frac{u^2}{2}\right) = u^2 \text{ for } 0 \le u \le 1$$

$$f_U(u) = \frac{dF_U(u)}{du}$$

$$= \begin{cases} 2u & 0 \le u \le 1 \\ 0 & \text{otherwise} \end{cases}$$

b. $E(U) = \int_0^1 u(2u)\, du = \frac{2}{3}$

c. $f_{X_i}(x_i) = \int_0^{1-x_i} 2dx_{i'} = 2(1-x_i)$ for $i, i' = 1, 2, \quad i \neq i', 0 \le x_i \le 1$

$$E(X_i) = \int_0^1 x_i [2(1-x_i)] dx_i$$

$$= 2\left(\frac{1}{2} - \frac{1}{3}\right) = \frac{1}{3} \quad i = 1, 2$$

$$E(X_1 + X_2) = E(X_1) + E(X_2)$$

$$= \frac{1}{3} + \frac{1}{3} = \frac{2}{3}$$

6.9 a. $U_1 = h_1(X) = 2X - 1 \Rightarrow X = h_1^{-1}(U) = \dfrac{U+1}{2}$

$$f_{U_1}(u) = f_X[h_1^{-1}(u)]\left|\frac{dx}{du}\right|$$

$$= 2\left(1 - \frac{u+1}{2}\right)\left|\frac{1}{2}\right| = \frac{1-u}{2} \text{ for } -1 \le u \le 1$$

b. $U_2 = h_2(X) = 1 - 2X \Rightarrow X = h_2^{-1}(U) = \dfrac{1-U}{2}$

$$f_{U_2}(u) = f_X[h_2^{-1}(u)]\left|\frac{dx}{du}\right|$$

$$= 2\left(1 - \frac{1-u}{2}\right)\left|-\frac{1}{2}\right|$$

$$= \frac{1+u}{2} \text{ for } -1 \le u \le 1$$

c. $U_3 = h_3(X) = X^2 \Rightarrow X = h_3^{-1}(U) = \sqrt{U}$

$$f_{U_3}(u) = f_X[h_3^{-1}(u)]\left|\frac{dx}{du}\right| = 2(1-\sqrt{u})\left|\frac{1}{2\sqrt{u}}\right|$$

$$= \frac{1-\sqrt{u}}{\sqrt{u}} \text{ for } 0 < u \le 1$$

6.11 Given $X_2 = x_2$, we have $U = h_{1|2}(X_1) = x_2X_1$, and $X_1|x_2 = h_{1|2}^{-1}(U) = \dfrac{U}{x_2}$, and

$$f_{U|x_2}(u|x_2) = f_{X_1|x_2}[h_{1|2}^{-1}(u)]\left|\frac{d(x_1|x_2)}{du}\right|$$

$$= 6\left[\frac{u}{x_2}\left(1 - \frac{u}{x_2}\right)\right]\left|\frac{1}{x_2}\right| = \frac{6u}{x_2^2}\left(1 - \frac{u}{x_2}\right) \text{ for } 0 < u \le x_2 \le 1$$

$$f_U(u) = \int_{-\infty}^{\infty} f(u|\,x_2) f(x_2)\, dx_2 = \int_u^1 \frac{6u}{x_2^2}\left(1 - \frac{u}{x^2}\right) 3x_2^2 dx_2 = \int_u^1 18u\left(1 - \frac{u}{x_2}\right) dx_2$$

$$= 18u\,\{\,(1-u) - u \ \ln \ x_2|_u^1\}$$

$$= 18u\,(1 - u + u \ \ln \ u) \text{ for } 0 < u \le 1.$$

6.13 **a.** $U = h(X) = X^m \Rightarrow X = h^{-1}(U) = U^{\frac{1}{m}}$. Thus

$$f_U(u) = f_X[h^{-1}(u)]\left|\frac{dx}{du}\right|$$

$$= \frac{1}{\alpha} m \left(u^{\frac{1}{m}}\right)^{m-1} e^{-\frac{(u^{1/m})^m}{\alpha}} \left|\frac{1}{m} u^{\frac{1}{m}-1}\right| = \frac{1}{\alpha} e^{-\frac{u}{\alpha}} \text{ for } u > 0.$$

b. Let $U = h(X) = X^k$. Then $X = h^{-1}(U) = U^{\frac{1}{k}}$, and

$$f_U(u) = f_X[h^{-1}(u)]\left|\frac{dx}{du}\right|$$

$$= \frac{1}{\alpha} m \left(u^{\frac{1}{k}}\right)^{m-1} e^{-\frac{(u^{1/k})^m}{\alpha}} \left|\frac{1}{k} u^{\frac{1}{k}-1}\right| = \frac{1}{\alpha}\left(\frac{m}{k}\right) u^{\frac{m}{k}-1} e^{-\frac{u^{m/k}}{\alpha}} \text{ for } u > 0$$

$$E(U) = \int_0^{\infty} \frac{1}{\alpha}\left(\frac{m}{k}\right) u^{m/k} e^{-\frac{u^{m/k}}{\alpha}} du = \int_0^{\infty} \frac{1}{\alpha}\left(\frac{m}{k}\right) v e^{-v/\alpha} \left(\frac{k}{m}\right) v^{\frac{k}{m}-1} dv$$

$$(\text{for } v = u^{m/k}) \quad = \int_0^{\infty} \frac{1}{\alpha} v^{\frac{k}{m}} e^{-\frac{v}{\alpha}} dv = \Gamma\left(\frac{k}{m} + 1\right) \alpha^{\frac{k}{m}}$$

6.15 **a.** Using the method of moment generating functions to find the new distribution, we have

$$M_Y(t) = E(e^{Yt}) = E(e^{\Sigma_{i=1}^k X_i t}) = E\left(\prod_{i=1}^{k} (e^{X_i t})\right)$$

$$* \qquad = \prod_{i=1}^{k} (E(e^{X_i t})) \text{ (since } X_i\text{'s are independent)}$$

$$= (M_X(t))^k \text{ (since } X_i\text{'s are identically distributed)}$$

$$= (e^{\lambda(e^t - 1)})^k = e^{k\lambda(e^t - 1)}$$

By the uniqueness of m.g.f.'s, then $Y \sim P(k\lambda)$.

b. In this case, each $X_i \sim P(\lambda_i)$. Starting from(*) above, we have

$$M_Y(t) = \prod_{i=1}^{k} (E(e^{X_i t})) = \prod_{i=1}^{k} (e^{\lambda_i (e^t - 1)}) = e^{\Sigma_{i=1}^{k} \lambda_i (e^t - 1)}$$

Therefore $Y \sim P\left(\sum_{i=1}^{k} \lambda_i\right)$

c. In this case, we are considering whether the distribution of $\Sigma a_i X_i$ for constants $a_1, \ldots, a_k$ is Poisson or not. Without loss of generality, let us consider the distribution of aX, $a \neq 0$, where $X \sim P(\lambda)$. Using the method of moment generating functions, we see that

$$M_Y(t) = E(e^{Yt}) = E(e^{aXt}) = E(e^{X(at)}) = e^{\lambda(e^{at} - 1)}$$

This is **not** the m.g.f. of a Poisson distribution, so aX is **not** distributed as a Poisson.

6.17 Using the method of moment generating functions, we have

$$M_{\bar{X}}(t) = E(e^{\bar{X}t}) = E\left(e^{\frac{1}{n}\Sigma_{i=1}^{n} X_i t}\right) = E\left(e^{\Sigma_{i=1}^{n} X_i \left(\frac{t}{n}\right)}\right) = E\left(\prod_{i=1}^{n} e^{X_i\left(\frac{t}{n}\right)}\right)$$

$$= \prod_{i=1}^{n} E\left(e^{X_i\left(\frac{t}{n}\right)}\right) \text{ (since } X_i\text{'s are independent)}$$

$$= \prod_{i=1}^{n} e^{\left(u\left(\frac{t}{n}\right) + \frac{\sigma^2}{2}\left(\frac{t}{n}\right)^2\right)} \text{ (since } X_i\text{'s are identically distributed)}$$

$$= \prod_{i=1}^{n} e^{\left(\frac{ut}{n} + \frac{\sigma^2 t^2}{2n^2}\right)} = \left(e^{\frac{ut}{n} + \frac{\sigma^2 t^2}{2n^2}}\right) = e^{ut + \frac{\sigma^2 t^2}{2n}}$$

By the uniqueness of the m.g.f., then $\bar{X} \sim N(\mu, \sigma^2/n)$.

6.19 It is known that the moment generating function of $(n-1)S^2/\sigma^2$ is

$$M_{\frac{(n-1)S^2}{\sigma^2}}(t) = (1-2t)^{-\left(\frac{n-1}{2}\right)}$$

Note that $M_{\frac{(n-1)S^2}{\sigma^2}}(t) = E\left(e^{\frac{(n-1)S^2}{\sigma^2} t}\right) = E\left(e^{S^2\left(\frac{(n-1)}{\sigma^2} t\right)}\right) = M_{S^2}\left(\frac{(n-1)}{\sigma^2} t\right)$

So the m.g.f. of S^2 is

$$M_{S^2}(t) = \left(1 - 2\,\frac{\sigma^2}{n-1}\,t\right)^{-\frac{n-1}{2}}$$

a. $E(S^2) = \frac{d}{dt}M_{S^2}(t)\Big|_{t=0} = \frac{-(n-1)}{2}\left(\frac{-2\sigma^2}{n-1}\right)\left(1 - 2\,\frac{\sigma^2}{n-1}\,t\right)^{-\frac{n-1}{2}-1}\Bigg|_{t=0}$

$$= \left(\frac{-(n-1)}{2}\right)\left(\frac{-2\sigma^2}{n-1}\right)(1) = \sigma^2$$

b. $V(S^2) = E(S^2)^2 - (E(S^2))^2 = E(S^2)^2 - (\sigma^2)^2$

and $E(S^2)^2 = \frac{d}{dt^2}M_{S^2}(t)\Big|_{t=0}$

$$= \frac{-(n-1)}{2}\left(\frac{-2\sigma^2}{n-1}\right)^2\left(\frac{-(n-1)}{2} - 1\right)\left(1 - 2\,\frac{\sigma^2}{n-1}\,t\right)^{-\frac{(n-1)}{2}-2}\Bigg|_{t=0}$$

$$= \frac{-(n-1)}{2}\left(\frac{4\sigma^4}{(n-1)^2}\right)\left(\frac{-(n-1)}{2} - 1\right)(1) = \sigma^4 + \frac{2\sigma^4}{n-1}$$

So $V(S^2) = \sigma^4 + \frac{2\sigma^4}{n-1} - \sigma^4 = \frac{2\sigma^4}{n-1}$

c. $E(S) = \int_0^\infty \sqrt{X}\frac{1}{\Gamma\left(\frac{n}{2}\right)2^{\frac{n}{2}}}x^{\frac{n}{2}-1}e^{-\frac{x}{2}}dx = \frac{1}{\Gamma\left(\frac{n}{2}\right)2^{\frac{n}{2}}}\int_0^\infty X^{\frac{n}{2}+\frac{1}{2}-1}e^{-\frac{x}{2}}dx$

$$= \frac{1}{\Gamma\left(\frac{n}{2}\right)2^{\frac{n}{2}}}\Gamma\left(\frac{n}{2} + \frac{1}{2}\right)2^{\frac{n}{2}+\frac{1}{2}} = \frac{\sqrt{2}\left(\Gamma\frac{n}{2} + \frac{1}{2}\right)}{\Gamma\left(\frac{n}{2}\right)}$$

6.21 **a.** $f_U(u) = g_1(u) = n[1 - F_X(u)]^{n-1}f_X(u) = 2[1-u]^{2-1}(1) = 2(1-u)$ for $0 \le u \le 1$

b. $f_U(u) = g_n(u) = n[F_X(u)]^{n-1}f_X(u) = 2(u)^{2-1} = 2u$ for $0 \le u \le 1$

6.23 **a.** $g_1(x) = n[1-F_X(x)]^{n-1}f_X(x) = n\left[1-\int_\theta^x e^{-(x-\theta)}dx\right]^{(n-1)} e^{-(x-\theta)}$

$$= n[1-(-e^{-(x-\theta)}|_\theta^x)]^{(n-1)} e^{-(x-\theta)} = n[1-(1-e^{-(x-\theta)})]^{(n-1)} e^{-(x-\theta)}$$

$$= ne^{-n(x-\theta)} \text{ for } x > \theta$$

b. $E(X_{(1)}) = \int_\theta^\infty xne^{-n(x-\theta)}dx = \int_0^\infty (y+\theta)ne^{-ny}dy = \int_0^\infty nye^{-ny}dy + \theta\int_0^\infty ne^{-ny}dy$

$$= \frac{1}{n}\int_0^\infty \frac{n^2}{\Gamma(2)}y^{2-1}e^{-ny}dy + \theta(-e^{-ny}|_0^\infty) = \frac{1}{n}+\theta$$

6.25 **a.** The density function of $X_{(1)}$ is given as

$$g_1(x) = n[1-F(x)]^{n-1}f(x)$$

where $F(x)$ and $f(x)$ are the distribution and density functions, respectively, of the $U(\theta_1, \theta_2)$ distribution. Hence,

$$g_1(x) = \begin{cases} n\left[1-\dfrac{x-\theta_1}{\theta_2-\theta_1}\right]^{n-1}\dfrac{1}{\theta_2-\theta_1} & \text{for } \theta_1 < x < \theta_2 \\ 0 & \text{elsewhere} \end{cases}$$

So, $E(X_{(1)}) = \int_{\theta_1}^{\theta_2} xn\left[1-\dfrac{x-\theta_1}{\theta_2-\theta_1}\right]^{n-1}\dfrac{1}{\theta_2-\theta_1}dx = n\int_{\theta_1}^{\theta_2}\dfrac{x}{\theta_2-\theta_1}\left[1-\dfrac{x-\theta_1}{\theta_2-\theta_1}\right]^{n-1}dx$

Let $y = \dfrac{x-\theta_1}{\theta_2-\theta_1}$. Then $x = \theta_1 + (\theta_2-\theta_1)y$ and $dx = (\theta_2-\theta_1)dy$ and

$$E(X_{(1)}) = n\int_0^1 \frac{\theta_1+(\theta_2-\theta_1)y}{\theta_2-\theta_1}[1-y]^{n-1}(\theta_2-\theta_1)dy$$

$$= n\int_0^1 \theta_1[1-y]^{n-1} + (\theta_2-\theta_1)y[1-y]^{n-1}dy$$

$$= \theta_1\int_0^1 n[1-y]^{n-1}dy + n(\theta_2-\theta_1)\int_0^1 y^{2-1}[1-y]^{n-1}dy$$

$$= -\theta_1[1-y]^n|_0^1 + n(\theta_2-\theta_1)\frac{\Gamma 2\Gamma n}{\Gamma(n+2)}$$

$$= \theta_1 + n(\theta_2-\theta_1)\frac{1!(n-1)!}{(n+1)!} = \theta_1 + \frac{\theta_2-\theta_1}{n+1} = \frac{\theta_2+n\theta_1}{n+1}$$

b. The density function of $X_{(n)}$ is given as

$$g_n(x) = n[F(x)]^{n-1}f(x)$$

where $F(x)$ and $f(x)$ as the distribution and density functions, respectively, of the $U(\theta_2, \theta_1)$ distribution. Hence,

$$g_n(x) = \begin{cases} n\left[\dfrac{x-\theta_1}{\theta_2-\theta_1}\right]^{n-1} \dfrac{1}{\theta_2-\theta_1} & \theta_1 < x < \theta_2 \\ 0 & \text{elsewhere} \end{cases}$$

$$\text{So } E(x_{(n)}) = \int_{\theta_1}^{\theta_2} xn\left[\frac{x-\theta_1}{\theta_2-\theta_1}\right]^{n-1} \frac{1}{\theta_2-\theta_1} dx = n\int_{\theta_1}^{\theta_2} \frac{x}{\theta_2-\theta_1}\left[\frac{x-\theta_1}{\theta_2-\theta_1}\right]^{n-1} dx$$

Let $y = \dfrac{x-\theta_1}{\theta_2-\theta_1}$. Then $x = \theta_1 + (\theta_2-\theta_1)y$ and $dx = (\theta_2-\theta_1)\,dy$ and

$$E(X_{(n)}) = n\int_0^1 \frac{\theta_1+(\theta_2-\theta_1)y}{(\theta_2-\theta_1)} (y)^{n-1}(\theta_2-\theta_1)\,dy$$

$$= n\int_0^1 \theta_1 y^{n-1} + (\theta_2-\theta_1)y^n dy = \theta_1\int_0^1 ny^{n-1}dy + n(\theta_2-\theta_1)\int_0^1 y^n dy$$

$$= \theta_1 y^n\Big|_0^1 + n(\theta_2-\theta_1)\frac{y^{n+1}}{n+1}\Big|_0^1 = \theta_1 + \frac{n(\theta_2-\theta_1)}{n+1}$$

$$= \frac{(n+1)\theta_1 + n\theta_2 - n\theta_1}{n+1} = \frac{n\theta_2+\theta_1}{n+1}$$

c. We want $E(h(X_{(1)}, X_{(n)})) = \theta_2 - \theta_1$. Consider $h(X_{(1)}, X_{(n)}) = bX_{(n)} - aX_{(1)}$. Then

$$E(h(X_{(1)}, X_{(n)})) = E(bX_{(n)} - aX_{(1)}) = bE(X_{(n)}) - aE(X_{(1)})$$

$$= b\frac{n\theta_2+\theta_1}{n+1} - a\frac{\theta_2+n\theta_1}{n+1} = \frac{bn\theta_2 + b\theta_1 - a\theta_2 - n\theta_1}{n+1}$$

$$= \frac{(bn-a)\theta_2 - (an-b)\theta_1}{n+1}.$$

So,

$$\frac{(bn-a)\theta_2 - (an-b)\theta_1}{n+1} = \theta_2 - \theta_1$$

$$\Rightarrow (bn-a)\theta_2 - (an-b)\theta_1 = (n+1)\theta_2 - (n+1)\theta_1$$

$$\Rightarrow bn - a = n+1 \qquad (1)$$

$$an - b = n+1 \qquad (2)$$

$$\Rightarrow a = (b-1)n - 1 \qquad (3)$$

Substituting (3) from (2) we have

$$[(b-1)n-1]n-b = n+1$$

$$\Rightarrow (b-1)n^2 - n - b = n+1$$

$$\Rightarrow bn^2 - n^2 - n - b = n+1$$

$$\Rightarrow b(n^2-1) = n^2+2n+1$$

$$\Rightarrow b = \frac{(n+1)^2}{(n-1)(n+1)}$$

$$\Rightarrow b = \frac{n+1}{n-1}$$

Substituting (2) into (1) gives us

$$bn - an - a + b = 0$$

$$\Rightarrow (b-a)n - (a-b) = 0$$

$$\Rightarrow (b-a)(n-1) = 0.$$

Since $n-1 = 0$ only if $n = 1$, then $b-a = 0$ for all n > 1. Hence,

$$a = \frac{n-1}{n+1}.$$

Therefore,

$$h(X_{(1)}, X_{(n)}) = \frac{n-1}{n+1}(X_{(n)} - X_{(1)}).$$

6.27 $R(s) = E(s^X) = \sum_{n=1}^{\infty} s^n \left(\frac{1}{n} \alpha q^n \right) = \alpha \sum_{n=1}^{\infty} \frac{(qs)^n}{n} = \alpha \ \ln \frac{1}{1-qs}$ for $|qs| < 1$

Then the probability generating function for this compound Poisson distribution is

$$P(s) = e^{-\lambda + \lambda R(s)} = e^{-\lambda + \lambda\alpha \ ln \frac{1}{1-qs}}$$

$$= e^{-\lambda} \left(\frac{1}{qs} \right)^{\alpha\lambda} = \left(\frac{e^{-\frac{1}{\alpha}}}{1-qs} \right) = \left(\frac{ps}{1-qs} \right) \text{ for } \alpha = -1/\ln \ ps$$

Let $Y_1, Y_2, \ldots, Y_r$ be independent geometric random variables with parameter p. Then $W = \Sigma_{i=1}^{r} Y_i$ has a negative binomial distribution with parameters r and p. The probability generating function for Y_i is $E(s^{Y_i}) = \frac{ps}{1-qs}$. Thus the probability generating function for W is

$$E(s^W) = E(s^{\Sigma Y_i}) = \prod_{i=1}^{r} E(s^{Y_i}) = \left(\frac{ps}{1-qs}\right)^r.$$

Therefore, $P(s) = E(s^W)$ for $\alpha = -1/\ln ps$ and $r = \alpha\lambda$.

6.29 **a.** $$P^{(2)}(s) = \sum_{n=2}^{\infty} s^n p_n^{(2)} = \sum_{n=2}^{\infty} s^n \sum_{\substack{i,j \\ i+j=n}} p_i^{(1)} p_j^{(1)} = \sum_{n=2}^{\infty} \sum_{\substack{i,j \\ i+j=n}} s^i p_i^{(1)} s^j p_j^{(1)}$$

$$= \sum_{i=1}^{\infty} \sum_{j=1}^{\infty} s^i p_i^{(1)} s^j p_j^{(1)} = \sum_{i=1}^{\infty} s^i p_i^{(1)} \sum_{j=1}^{\infty} s^j p_j^{(1)} = P^2(s)$$

b. $$P(s) = \sum_{n=1}^{\infty} s^n p_n^{(1)} = s p_1^{(1)} + \sum_{n=2}^{\infty} s^n p_n^{(1)} = sp + qs \sum_{n=2}^{\infty} s^{n-1} p_{n-1}^{(2)}$$

$$= sp + qs \sum_{i=1}^{\infty} s^i p_i^{(2)} = sp + qsP^{(2)}(s) = sp + qsP^2(s)$$

c. By the Quadratic Equation,

$$P(s) = \frac{1 \pm \sqrt{1 - 4qs(ps)}}{2qs}$$

i.e.,

$$\frac{1 - \sqrt{1 - 4qps^2}}{2qs}$$

is a solution for $P(s)$.

d. $$P(1) = \frac{1 - \sqrt{1 - 4pq(1^2)}}{2q(1)} = \frac{1 - \sqrt{1 - 4pq}}{2q} = \frac{1 - \sqrt{1 - 4p + 4p^2}}{2q}$$

$$= \frac{1 - \sqrt{[p - (1-p)]^2}}{2q} = \frac{1 - \sqrt{(p-q)^2}}{2q} = \frac{1 - |p-q|}{2q} = \frac{1 - |p-q|}{1 - (1-2q)}$$

$$= \frac{1 - |p-q|}{1 - (p+q-2q)} = \frac{1 - |p-q|}{1 - (p-q)} = 1 \text{ for } p > q$$

$$= \frac{1 - q + p}{1 - p + q} = \frac{p}{q} \text{ for } p \le q$$

i.e., $P(1) = \Sigma_{n=1}^{\infty} 1^n p_n^{(1)} = \Sigma_{n=1}^{\infty} p_n^{(1)} =$ probability that S_n ever becomes positive

$$= \begin{cases} 1 \text{ for } p > q \\ \frac{p}{q} \text{ for } p \le q \end{cases}$$

e. $P(s) = \dfrac{1-\sqrt{1-4pqs^2}}{2qs}$

$$= \frac{1-\sqrt{1-4\left(\frac{1}{2}\right)\left(\frac{1}{2}\right)s^2}}{2\left(\frac{1}{2}\right)s} = \frac{1-\sqrt{1-s^2}}{s} = \frac{1}{x} - \sqrt{\frac{1}{s^2}-1}$$

E(number of trials until the first passage of S_N through 1) =

$$P'(1) = \left\{-\frac{1}{s^2} - \frac{1}{2\sqrt{\frac{1}{s^2}-1}}\left(\frac{-2}{s^3}\right)\right\}\Bigg|_{s=1}$$

$$= -1 + \infty = \infty$$

6.31 **a.** $U = X_1 + X_2$ is distributed as a Poisson with mean $\lambda_1 + \lambda_2$.

b. $P(X_1 = u \mid X_1 + X_2 = m) = \dfrac{P(X_1 = u, X_1 + X_2 = m)}{P(X_1 + X_2 = m)} = \dfrac{P(X_1 = u, X_2 = m-u)}{P(X_1 + X_2 = m)}$

$$= \frac{\dfrac{e^{-\lambda_1}\lambda_1^u}{u!}\cdot\dfrac{e^{-\lambda_2}\lambda_2^{m-u}}{(m-u)!}}{\dfrac{e^{-(\lambda_1+\lambda_2)}(\lambda_1+\lambda_2)^m}{m!}} = \frac{m!}{u!\,(m-u)!}\cdot\frac{\lambda_1^u\lambda_2^{m-u}}{(\lambda_1+\lambda_2)^m}$$

$$= \binom{m}{u}\left(\frac{\lambda_1}{\lambda_1+\lambda_2}\right)^u\left(\frac{\lambda_2}{\lambda_1+\lambda_2}\right)^{m-u}$$

i.e., $X_1 \mid (X_1 + X_2 = m)$ has a binomial distribution with parameters m, and $\dfrac{\lambda_1}{\lambda_1+\lambda_2} = p$.

6.33 Let $U = |X|$. Then

$$F_U(u) = P(U \le u) = P(|X| \le u) = P(-u \le X \le u)$$

$$= F_X(u) - F_X(-u) = \int_{-1}^{u}\frac{1}{2}dx - \int_{-1}^{-u}\frac{1}{2}dx$$

$$= \frac{1}{2}[u+1-(-u+1)] = u \text{ for } 0 \le u \le 1$$

$$f_U(u) = \frac{dF_U(u)}{du}$$

$$= \begin{cases} 1 & 0 \le u \le 1 \\ 0 & \text{otherwise} \end{cases}$$

6.35 Given $X_1 = x_1$, we have $U = h_{2|1}(X_2) = \frac{1}{2}(x_1 + X_2)$, and $X_2|x_1 = h_{2|1}^{-1}(U) = 2U - x_1$, and

$$f_{U|x_1}(u|x_1) = f_{X_2|x_1}[h_{2|1}^{-1}(u)]\left|\frac{d(x_2|x_1)}{du}\right| = f_{X_2}(2u - x_1)\,|2|$$

$$= \frac{1}{2}(2u - x_1)\,e^{-\frac{2u - x_1}{2}}$$

$$f_U(u) = \int_{-\infty}^{\infty} f(u|x_1)f(x_1)\,dx_1 = \int_0^{2u} \frac{1}{2}(2u - x_1)\,e^{-\frac{2u - x_1}{2}}\left(\frac{1}{4}x_1 e^{-\frac{x_1}{2}}\right)dx_1$$

$$= \int_0^{2u} \frac{1}{8}x_1(2u - x_1)\,e^{-u}dx_1 = \frac{e^{-u}}{8}\left\{2u\left(\frac{4u^2}{2}\right) - \frac{8u^3}{3}\right\} = \left(\frac{e^{-u}u^3}{6} = \frac{e^{-u}u^3}{\Gamma(4)}\right)$$

i.e., U has a Gamma distribution with parameters $\alpha = 4, \beta = 1$.

6.37 Let X_i = distance of assigned post i, $i = 1, 2$. Then X_1 and X_2 are independent uniform random variables with density function $f(x) = 1$. Let $U = |X_1 - X_2|$. Then

$$F_U(u) = P(|X_1 - X_2| \le u) = P(-u \le X_1 - X_2 \le u)$$

$$= P(X_2 - u \le X_1 \le X_2 + u)$$

$$= \int_0^u \int_0^{x_1 + u} dx_2 dx_1 + \int_u^{1-u} \int_{x_1 - u}^{x_1 + u} dx_2 dx_1 + \int_{1-u}^1 \int_{x_1 - u}^1 dx_2 dx_1$$

$$= \int_0^u (x_1 + u)\,dx_1 + \int_u^{1-u} 2u\,dx_1 + \int_{1-u}^1 (1 - x_1 + u)\,dx_1$$

$$= \frac{u^2}{2} + u^2 + 2u(1 - 2u) + (1 - 1 + u) - \frac{1}{2}[1 - (1-u)^2] + u^2$$

$$= \left(\frac{u^2}{2} + u^2 - 4u^2 + \frac{u^2}{2} + u^2\right) + 2u + u - u = 2u - u^2 \text{ for } 0 \le u \le 1$$

probability region of interest:

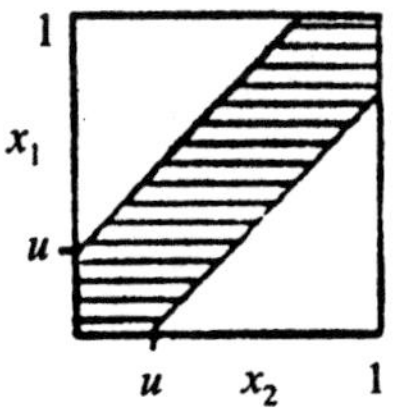

$$P\left(U \le \frac{1}{2}\right) = F_U\left(\frac{1}{2}\right) = -\left(\frac{1}{2}\right)^2 + 2\left(\frac{1}{2}\right) = 1 - \frac{1}{4} = \frac{3}{4}$$

6.39 For $0 \le u \le 1$:

$$f_U(u) = \frac{1}{2\sqrt{u}}[f_X(\sqrt{u}) - f_X(-\sqrt{u})] = \frac{1}{2\sqrt{u}}\left(\frac{1}{4} + \frac{1}{4}\right) = \frac{1}{4\sqrt{u}}$$

for $1 < X \le 3$; i.e., $1 < u \le 9$:

$$f_U(u) = f_X(\sqrt{u})\left|\frac{dx}{du}\right| = f_X(\sqrt{u})\frac{1}{2\sqrt{u}} = \frac{1}{4}\left(\frac{1}{2\sqrt{u}}\right) = \frac{1}{8\sqrt{u}}$$

6.41 $$g_{12\ldots n}(x_1, \ldots, x_n) = n! f(x_1)\ldots f(x_n)\; x_1 \le x_2 \le \ldots \le x_n$$

where $f(x_i) = 1, 0 \le x_i \le 1, i = 1, \ldots, n$; i.e., $g_{12\ldots n}(x_1, \ldots, x_n) = n!$, and

$$g_{1n}(x_1, x_n) = n!\int_{x_1}^{x_n}\int_{x_1}^{x_{n-1}}\ldots\int_{x_1}^{x_3} dx_2 dx_3 \ldots dx_{n-1}$$

$$= n!\frac{(x_n - x_1)^{n-2}}{(n-2)!} = n(n-1)(x_n - x_1)^{n-2}$$

Let $R = X_{(n)} - X_{(1)}$. Then

$$F_R(u) = P(R \le u) = PX_{(n)} - (X_{(1)} \le u) = P(X_{(n)} \le u + X_{(1)})$$

$$= 1 - \int_u^1\int_0^{x_n - u} n(n-1)(x_n - x_1)^{n-2} dx_1 dx_n$$

$$= 1 + \int_u^1 n[(x_n - x_1)^{n-1}\big|_0^{x_n - u}]\,dx_n = 1 + \int_u^1 n(u^{n-1} - x_n^{n-1})\,dx_n$$

$$= 1 + nu^{n-1}(1-u) - x_n^n\big|_u^1 = nu^{n-1} - u^n(n-1)$$

probability region of interest:

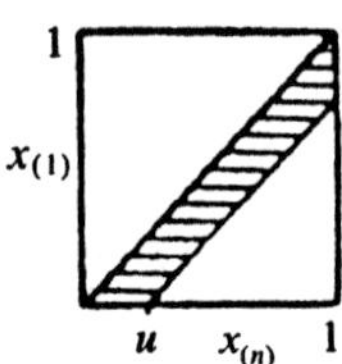

$$f_R(u) = \frac{dF_R(u)}{du}$$

$$= \begin{cases} n(n-1)(u^{n-2} - u^{n-1}) & 0 \le u \le 1 \\ 0 & \text{otherwise} \end{cases}$$

6.43 **a.** $F_{F|w}(f|w) = P(F \le f|w) = P(V \le cf|w)$

$$= \int_0^{cf} \frac{1}{\Gamma\left(\frac{n_1}{2}\right)} \left(\frac{1}{2}\right)^{\frac{n_1}{2}} y^{\frac{n_1}{2}-1} e^{-\frac{y}{2}} dy = \int_0^{f} \frac{1}{\Gamma\left(\frac{n_1}{2}\right)} \left(\frac{1}{2}\right)^{\frac{n_1}{2}} (cx)^{\frac{n_1}{2}-1} e^{-\frac{cx}{2}} c\,dx$$

$$f_{F|w}(f|w) = \frac{1}{\Gamma\left(\frac{n_1}{2}\right)} \left(\frac{c}{2}\right)^{\frac{n_1}{2}} f^{\frac{n_1}{2}-1} e^{-\frac{cf}{2}} = \frac{1}{\Gamma\left(\frac{n_1}{2}\right)} \left(\frac{wn_1}{2n_2}\right)^{\frac{n_1}{2}} f^{\frac{n_1}{2}-1} e^{-f\left(\frac{wn_1}{2n_2}\right)}.$$

b. $f(f, w) = f(f|w)f(w)$

$$= \frac{1}{\Gamma\left(\frac{n_1}{2}\right)} \left(\frac{wn_1}{2n_2}\right)^{\frac{n_1}{2}} f^{\frac{n_1}{2}-1} e^{-f\left(\frac{wn_1}{2n_2}\right)} \frac{1}{\Gamma\left(\frac{n_2}{2}\right)} \left(\frac{1}{2}\right)^{\frac{n_2}{2}} w^{\frac{n_2}{2}-1} e^{-\frac{w}{2}}$$

$$= \frac{1}{\Gamma\left(\frac{n_1}{2}\right)\Gamma\left(\frac{n_2}{2}\right)} \left(\frac{1}{2}\right)^{\frac{n_1+n_2}{2}} \left(\frac{n_1}{n_2}\right)^{\frac{n_1}{2}} f^{\frac{n_1}{2}-1} w^{\frac{n_1+n_2}{2}-1} e^{-\frac{w}{2}\left(\frac{fn_1}{n_2}+1\right)}$$

c. $$g(f) = \frac{1}{\Gamma\left(\frac{n_1}{2}\right)\Gamma\left(\frac{n_2}{2}\right)} \left(\frac{1}{2}\right)^{\frac{n_1+n_2}{2}} \left(\frac{n_1}{n_2}\right)^{\frac{n_1}{2}} f^{\frac{n_1}{2}-1} \int_0^{\infty} w^{\frac{n_1+n_2}{2}-1} e^{-\frac{w}{2}\left(\frac{fn_1}{n_2}+1\right)} dw$$

$$= \frac{\Gamma\left(\frac{n_1+n_2}{2}\right)}{\Gamma\left(\frac{n_1}{2}\right)\Gamma\left(\frac{n_2}{2}\right)} \left(\frac{n_1}{n_2}\right)^{\frac{n_1}{2}} f^{\frac{n_1}{2}-1} \left(\frac{1}{\frac{fn_1}{n_2}+1}\right)^{\frac{n_1+n_2}{2}}$$

6.45 **a.** Let X_i = life length of ith component functioning at time W_{j-1} $i = 1, 2, \ldots, (n-j+1)$. Recall that $P(X_i > b | X_i > a) = P(X_i > b - a)$, for $b > a$. Then T_j = length of time between the $(j-1)^{th}$ failure and the jth failure = $\min(X_1, \ldots, X_{n-j+1})$, and

$$f_{T_j}(t) = (n-j+1)\,[1 - F_{X_i}(t)]^{(n-j+1)-1} f_{X_i}(t)$$

$$= (n-j+1)\,(e^{-t/\theta})^{n-j}\left(\frac{1}{\theta}\right)e^{-t/\theta}$$

$$= \left(\frac{n-j+1}{\theta} \right) e^{-\left(\frac{n-j+1}{\theta}\right) t}$$

i.e., T_j has an exponential distribution with mean $\dfrac{\theta}{n-j+1}$.

b. $$U_r = \sum_{j=1}^{r} (n-j+1) T_j = (n+1) \sum_{j=1}^{r} T_j - \sum_{j=1}^{r} j T_j$$

$$= (n+1) \left[W_1 + \sum_{j=2}^{r} (W_j - W_{j-1}) \right] - \left[W_1 + \sum_{j=2}^{r} j (W_j - W_{j-1}) \right]$$

$$= (n+1) \left[W_1 + (W_2 - W_1) + (W_3 - W_2) + \ldots + (W_r - W_{r-1}) \right]$$

$$- \left[W_1 + 2 (W_2 - W_1) + 3 (W_3 - W_2) + \ldots + r (W_r - W_{r-1}) \right]$$

$$= (n+1) W_r - \left[r W_r - \sum_{j=1}^{r-1} W_j \right] = \sum_{j=1}^{r} W_j + (n-r) W_r$$

and

$$E(U_r) = \sum_{j=1}^{r} (n-j+1) E(T_j) = \sum_{j=1}^{r} (n-j+1) \frac{\theta}{(n-j+1)} = r\theta$$

6.47 $$E = h(v) = \frac{1}{2} m V^2$$

$$V = h^{-1}(E) = \sqrt{\frac{2E}{m}}$$

$$f_E(u) = f_V[h^{-1}(u)] \left| \frac{dv}{de} \right| = \frac{4}{\sqrt{\pi}} b^{3/2} \frac{2u}{m} e^{-\frac{b2u}{m}} \left| \frac{1}{\sqrt{2mu}} \right|$$

$$= 4 \sqrt{\frac{2}{\pi}} \left(\frac{b}{m} \right)^{3/2} \sqrt{u} e^{-\frac{2bu}{m}} \text{ for } u > 0$$

CHAPTER 7
SOME APPROXIMATIONS TO PROBABILITY DISTRIBUTIONS: LIMIT THEOREMS

7.1 By Theorem 7.1, $\bar{Y}_n$ converges in probability to μ, where

$$\mu = E(Y_i) = \int_0^1 y(3y^2)\,dy = \frac{3}{4}.$$

7.3 **a.** By Theorem 7.1, $\bar{Y}_n$ converges in probability to μ, where

$$\mu = E(Y_i) = \frac{\theta}{2}.$$

b. Let $Y_{(n)} = \max(Y_1, \ldots, Y_n)$. From Example 7.3 we have that

$$\lim_{n \to \infty} P(Y_{(n)} = \theta) = 1.$$

Hence, $\lim_{n \to \infty} P(|Y_{(n)} - \theta| < \varepsilon) = 1$ for every positive number ε.

7.5 Let X_n = number of persons that suffer a bad reaction out of a sample of n. Then X_n has a binomial distribution with parameters n, and $p = 0.001$, and from Example 7.4 we have that X_n converges in distribution to a Poisson with parameter $\lambda = np = (0.001)n$. Therefore, $n = 1{,}000$,

$$P(X_n > 2) = 1 - P(X_n \le 1) = 1 - \sum_{x=0}^{1} \frac{[(0.001)n]^x e^{(0.001)n}}{x!} = 1 - 2e^{-1}$$

7.7 Let $\bar{X}$ = sample mean

$$P(|\bar{X} - \mu| \le 1) = P(-1 \le \bar{X} - \mu \le 1) = P\left(\frac{-\sqrt{n}}{\sigma} \le \frac{\sqrt{n}(\bar{X} - \mu)}{\sigma} \le \frac{\sqrt{n}}{\sigma}\right)$$

$$\approx P(-\frac{\sqrt{100}}{10} \le Z \le \frac{\sqrt{100}}{10}) = P(-1 \le Z \le 1)$$

$$= 2(0.3413) = 0.6826$$

7.9 Let $\bar{X}$ = sample mean. Since, for mound-shaped symmetric distributions 95% of the observations lie within 2 standard deviations of the mean, assume the standard deviation of the pH to be one fourth of the usual range; i.e., $\sigma = (8-5)/4 = 3/4$. Then

$$P(|\bar{X}-\mu| \le 0.2) = P\left(-\frac{0.2\sqrt{40}}{(3/4)} \le \frac{(\bar{X}-\mu)\sqrt{40}}{(3/4)} \le \frac{0.2\sqrt{40}}{(3/4)}\right)$$

$$\approx P(-1.687 \le Z \le 1.687) = 2(0.4545) = 0.9090$$

7.11 **a.** $$P(199 < \bar{X} < 202) = P\left(\frac{\sqrt{25}(199-200)}{10} < \frac{\sqrt{n}(\bar{X}-\mu)}{\sigma} < \frac{\sqrt{25}(202-200)}{10}\right)$$

$$\approx P(-0.5 < Z < 1) = 0.1915 + 0.3413 = 0.5328$$

b. $$P\left(\sum_{i=1}^{25} X_i \le 5100\right) = P\left(\frac{\Sigma_{i=1}^{25} X_i}{25} \le \frac{5100}{25}\right) = P(\bar{X} \le 204)$$

$$= P\left(\frac{\sqrt{n}(\bar{X}-\mu)}{\sigma} \le \frac{\sqrt{25}(204-200)}{10}\right)$$

$$\approx P(Z \le 2) = 0.5 + 0.4772 = 0.9772$$

c. The approximations in parts (a) and (b) assume that the resistances are independent, and random sampling.

7.13 $$P(\bar{X} < 1.3) = P\left(\frac{\sqrt{n}(\bar{X}-\mu)}{\sigma} < \frac{\sqrt{25}(1.3-1.4)}{0.05}\right)$$

$$\approx P(Z < -10) \approx 0$$

7.15 **a.** $$P\left(\sum_{i=1}^{100} X_i > 45\right) = P(\bar{X} > 0.45)$$

$$\approx P\left(Z > \frac{\sqrt{100}(0.45-0.5)}{0.2}\right)$$

$$= P(Z > -2.5) = 0.5 + 0.4938 = 0.9938$$

b. $P\left(\sum_{i=1}^{n} X_i > 50\right) = P\left(\bar{X} > \frac{50}{n}\right)$

$$\approx P\left(Z > \frac{\sqrt{n}\left(\frac{50}{n} - 0.5\right)}{0.2}\right) = 0.99$$

This equation is true for

$$\frac{\sqrt{n}\left(\frac{50}{n} - 0.5\right)}{0.2} = -2.33$$

i.e.,

$$0.5n - 0.466\sqrt{n}\,(-50) = 0.$$

Using the quadratic formula, we have

$$\sqrt{n} = \frac{0.466 \pm \sqrt{(0.466)^2 - 4\,(0.5)\,(-50)}}{2\,(0.5)} = -9.5449 \text{ or } 10.4769.$$

Using the positive root, we have $n = (10.4769)^2 = 109.77$; i.e., $n = 110$.

7.17 Let X_i = service time for i^{th} customer, $i = 1, \ldots, 100$.

$$P\left(\sum_{i=1}^{100} X_i < 120\right) = P\left(\bar{X} < \frac{120}{100}\right) = P(\bar{X} < 1.2)$$

$$= P\left(\frac{\sqrt{n}\,(\bar{X} - \mu)}{\sigma} < \frac{\sqrt{100}\,(1.2 - 1.5)}{1}\right)$$

$$\approx P(Z < -3) = 0.5 - 0.4987 = 0.0013$$

7.19 $P(|\bar{X} - \bar{Y} - (\mu_1 - \mu_2)| \le 0.05) = P\left(\left|\frac{(\bar{X} - \bar{Y} - (\mu_1 - \mu_2))}{\sqrt{\frac{\sigma_1^2}{n_1} + \frac{\sigma_2^2}{n_2}}}\right| \le \frac{0.05}{\sqrt{\frac{0.01}{50} + \frac{0.02}{100}}}\right)$

$$\approx P(|Z| \le 2.5) = 2P(0 \le Z \le 2.5)$$

$$= 2\,(0.4938) = 0.9876$$

7.21 Let $\overline{X}$ and $\overline{Y}$ be the sample means for A and B, respectively. Then

$$P(\overline{Y}-\overline{X}>1) = P\left(\frac{\overline{Y}-\overline{X}-0}{\sqrt{\frac{\sigma^2}{n}+\frac{\sigma^2}{n}}} > \frac{1-0}{\sqrt{2\left(\frac{4}{50}\right)}}\right)$$

$$\approx P(Z>2.5) = 0.5-0.4938 = 0.0062$$

7.23 Let Y = number of nonconformances out of the sample of 50. Then Y has a binomial distribution with parameters $n = 50, p$. Let X be a normally distributed random variable with parameters $\mu = np = 50p$, $\sigma^2 = np(1-p) = 50p(1-p)$. Then

$$P(\text{accepting the lot}) = P(Y\leq 5)$$

$$\approx P(X\leq 5.5)$$

$$= P\left(\frac{X-\mu}{\sigma} \leq \frac{5.5-50p}{\sqrt{50p(1-p)}}\right) = P\left(Z\leq \frac{5.5-50p}{\sqrt{50p(1-p)}}\right)$$

a. For $p = 0.1$, we have

$$P(Y\leq 5) \approx P\left(Z\leq \frac{5.5-50(0.1)}{\sqrt{50(0.1)(1-0.1)}}\right)$$

$$= P(Z\leq 0.24) = 0.5948.$$

b. For $p = 0.2$, we have

$$P(Y\leq 5) \approx P\left(Z\leq \frac{5.5-50(0.2)}{\sqrt{50(0.2)(1-0.2)}}\right)$$

$$= P(Z\leq -1.59) = 0.5-0.4441 = 0.0559.$$

c. For $p = 0.3$, we have

$$P(Y\leq 5) \approx P\left(Z\leq \frac{5.5-50(0.3)}{\sqrt{50(0.3)(1-0.3)}}\right)$$

$$= P(Z\leq -2.93) = 0.5-0.4983 = 0.0017.$$

7.25 Let Y = number of disks that contain missing pulses out of the sample of 100. Then Y has a binomial distribution with parameters $n = 100$, $p = 0.2$. Let X be a normally distributed random variable with parameters $\mu = np = 100(0.2) = 20$, $\sigma^2 = np(1-p) = 100(0.2)(0.8) = 16$. Then

$$P(Y\leq 15) \approx P(X\leq 15.5)$$

$$= P\left(\frac{X-\mu}{\sigma} \leq \frac{15.5-20}{\sqrt{16}}\right)$$

$$= P(Z\leq -1.125) = 0.5-0.3708 = 0.1292$$

7.27 Let Y = number of days in which demand for over 500,000 gallons per day occurred, out of the sample of 30. Then Y has a binomial distribution with parameters $n = 30$, $p = 0.15$. Let X be a normally distributed random variable with parameters $\mu = np = 30(0.15) = 4.5$, $\sigma^2 = np(1-p) = 30(0.15)(0.85) = 3.825$. Then

$$P(Y \le 2) \approx P(X \le 2.5)$$

$$= P\left(\frac{X-\mu}{\sigma} \le \frac{2.5-4.5}{\sqrt{3.825}}\right) = P(Z \le -1.02) = 0.5 - 0.3461 = 0.1539$$

7.29 Let W = waiting time. Then

$$P(W > 10) = 1 - F(10) = 1 - \left(1 - e^{-\frac{10}{10}}\right) = e^{-1}$$

Let Y = number of customers whose waiting time exceeds 10 minutes, out of the sample of 100. Then Y has a binomial distribution with parameters $n = 100$, $p = P(W > 10) = e^{-1}$. Let X be a normally distributed random variable with parameters $\mu = np = 100e^{-1} = 36.7880$, $\sigma^2 = np(1-p) = 100e^{-1}(1-e^{-1}) = 23.2544$. Then

$$P(Y \ge 50) \approx P(X \ge 49.5)$$

$$= P\left(\frac{X-\mu}{\sigma} \ge \frac{49.5-36.7880}{\sqrt{23.2544}}\right) = P(Z \ge 2.6361) = 0.5 - 0.4959 = 0.0041$$

7.31 Let Y = number of vouchers that show up as being improperly documented out of the sample of 100. Then Y has a binomial distribution with parameters $n = 100$, $p = 0.2$. Let X be a normally distributed random variable with parameters $\mu = np = 100(0.2) = 20$, $\sigma^2 = np(1-p) = 100(0.2)(0.8) = 16$.

$$P(Y > 30) \approx P(X > 30.5)$$

$$= P\left(\frac{X-\mu}{\sigma} > \frac{30.5-20}{\sqrt{16}}\right) = P(Z > 2.63) = 0.5 - 0.4957 = 0.0043$$

7.33 $$P\left(|\bar{X}-\mu| \le \frac{1}{2}\right) = P\left(-\frac{1}{2} \le \bar{X}-\mu \le \frac{1}{2}\right)$$

$$= P\left(\frac{\sqrt{100}\,(0.5)}{2.5} \le \frac{\sqrt{n}\,(\bar{X}-\mu)}{\sigma} \le \frac{\sqrt{100}\,(0.5)}{2.5}\right)$$

$$\approx P(-2 \le Z \le 2) = 2\,(0.4772) = 0.9544$$

7.35 Let Y = number of defectives out of a sample of 100. Then Y has a binomial distribution with parameters $n = 100, p = 0.1$. Let X be a normally distributed random variable with parameters $\mu = np = 100(0.1) = 10, \sigma^2 = np(1-p) = 100(0.1)(0.9) = 9$.

$$P(Y \geq 15) \approx P(X \geq 14.5)$$

$$= P\left(\frac{X-\mu}{\sigma} \geq \frac{14.5-10}{\sqrt{9}}\right) = P(Z \geq 1.5) = 0.5 - 0.4332 = 0.0668$$

7.37 Let X_i = life of heat lamp i, $E(X_i) = \mu = 50$, $V(X_i) = \sigma^2 = (4)^2$. Let N_t = number of failed heat lamps at time t. Then N_t is approximately distributed as a normal random variable with mean $\frac{t}{\mu} = \frac{1,300}{50} = 26$, and variance $\frac{t\sigma^2}{\mu^3} = \frac{1,300(16)}{(50)^3} = 0.1664$. Hence

$$P(N_t < 25) = P\left(\frac{N_t - \frac{t}{\mu}}{\sqrt{\frac{t\sigma^2}{\mu^3}}} < \frac{25-26}{\sqrt{0.1664}}\right)$$

$$\approx P(Z < -2.45) = 0.5 - 0.4929 = 0.0071$$

7.39 By Theorem 7.1, $\bar{X}$ converges in probability to λ_1, and $\bar{Y}$ converges in probability to λ_2. By Theorem 7.2 (a), $\bar{X}+\bar{Y}$ converges in probability to $\lambda_1+\lambda_2$. Therefore, by Theorem 7.2 (c),

$$\frac{\bar{X}}{\bar{X}+\bar{Y}}$$

converges in probability to

$$\frac{\lambda_1}{\lambda_1+\lambda_2}.$$

7.41 $$P\left(\sum_{i=1}^{50} C_i > 48\right) = P\left(\sum_{i=1}^{50} 4(Y_i-\mu)^2 > 48\right) = P\left(\sum_{i=1}^{50} (Y_i-\mu)^2 > 12\right)$$

$$= P\left(\sum_{i=1}^{50} \frac{(Y_i-\mu)^2}{0.2} > \frac{12}{0.2}\right)$$

$$\approx P\left(\sum_{i=1}^{50} \chi_1^2 > 60\right)$$

$$= P(\chi_{50}^2 > 60)$$

$$= P\left(\frac{\chi_{50}^2 - 50}{\sqrt{2(50)}} > \frac{60-50}{\sqrt{100}}\right)$$

$$\approx P(Z > 1) \text{ (using the result of Exercise 7.40)}$$

$$= 0.5 - 0.3413 = 0.1587$$

CHAPTER 8
EXTENDED APPLICATIONS OF PROBABILITY

8.1 **a.** Let $Y(t)$ = number of calls in the next t minutes. If the number of calls in one period of time is independent of the number of calls in another non-overlapping time period, then $Y(t)$ has a Poisson distribution with a mean of $\left(\frac{1}{3}\right)t$, and

$$P[Y(5) \le 1] = P[Y(5) = 0] + P[Y(5) = 1] = P_0(5) + P_1(5) = e^{-(5)\,1/3} + 5\left(\frac{1}{3}\right)$$

$$e^{-(5)\,1/3} = e^{-5/3} + \frac{5}{3}e^{-5/3} = 0.5037.$$

b. $P[Y(5) = 2 | Y(2) = 2] = P[Y(3) = 0]$

since we assume the number of calls in two non-overlapping time intervals are independent, and

$$P[Y(3) = 0] = e^{-\frac{1}{3}(3)} = e^{-1} = 0.3679$$

8.3 Let $P_n = \left(1 - \frac{\lambda}{\theta}\right)\left(\frac{\lambda}{\theta}\right)^n$; i.e., P_n is the geometric equilibrium distribution. Then

$$\lambda P_{n-1} - (\lambda + \theta)P_n + \theta P_{n+1}$$

$$= \lambda\left(1 - \frac{\lambda}{\theta}\right)\left(\frac{\lambda}{\theta}\right)^{n-1} - (\lambda + \theta)\left(1 - \frac{\lambda}{\theta}\right)\left(\frac{\lambda}{\theta}\right)^{n} + \theta\left(1 - \frac{\lambda}{\theta}\right)\left(\frac{\lambda}{\theta}\right)^{n+1}$$

$$= \left(1 - \frac{\lambda}{\theta}\right)\left(\frac{\lambda}{\theta}\right)^{n-1}\left\{\lambda - (\lambda + \theta)\left(\frac{\lambda}{\theta}\right) + \theta\left(\frac{\lambda}{\theta}\right)\right\}$$

$$= \left(1 - \frac{\lambda}{\theta}\right)\left(\frac{\lambda}{\theta}\right)^{n-1}\left\{\lambda - \frac{\lambda^2}{\theta} - \lambda + \frac{\lambda^2}{\theta}\right\}$$

$$= 0$$

i.e., the geometric equilibrium distribution is a solution to the differential equation defining the equilibrium state; i.e.,

$$\lambda P_{n-1} - (\lambda + \theta)P_n + \theta P_{n+1} = 0.$$

8.5 Note that $Y(t)$ and $X(t)$ each have a Poisson distribution. Let $\lambda_1 t$ = mean of $Y(t)$ and $\lambda_2 t$ = mean of $X(t)$. Then, $Y(t) + X(t)$ has a Poisson distribution with mean $(\lambda_1 + \lambda_2)\,t$.

8.7 Let $U_{(i)}$ = time the ith call arrived, $i = 1, 2$, and $T = U_{(2)} - U_{(1)}$. Then

$$P(T > d) = P(U_{(2)} - U_{(1)} > d) = P(U_{(2)} > d + U_{(1)})$$

$$= \int_0^{t-d}\int_{d+u_{(1)}}^{t} g(u_1, u_2)\,du_2 du_1 = \int_0^{t-d}\int_{d+u_{(1)}}^{t} \frac{2}{t^2} du_2 du_1$$

$$= \int_0^{t-d} \frac{2}{t^2}(t - d - u_1)\,du_1 = \frac{2}{t^2}(t-d)^2 - \frac{1}{t^2}(t-d)^2 = \left(1 - \frac{d}{t}\right)^2$$

8.9 **a.** $F_R(r) = P(R \le r) = 1 - P(R > r) = 1 - P\left(\frac{4\pi R^3}{3} > \frac{4\pi r^3}{3}\right)$

$$= 1 - P\left[Y\left(\frac{4\pi r^3}{3}\right) \quad 0\right] = 1 - e^{-\frac{4\pi r^3 \lambda}{3}}$$

$$f_R(r) = \frac{\partial F_R(r)}{\partial r}$$

$$= \begin{cases} 4\lambda\pi r^2 e^{-\frac{4\pi r^3\lambda}{3}} & r > 0 \\ 0 & \text{otherwise} \end{cases}$$

b. $F_U(u) = P(R^3 \le u) = P(R \le u^{1/3})$

$$= \int_0^{u^{1/3}} 4\lambda\pi r^2 e^{-\frac{4\pi r^3\lambda}{3}} dr = -e^{-\frac{4\pi r^3\lambda}{3}}\Bigg|_0^{u^{1/3}} = 1 - e^{-\frac{4\pi\lambda u}{3}} \quad u > 0$$

i.e., U has an exponential distribution with mean $3/(4\lambda\pi)$.

8.11 **a.** The expected cost per unit time is

$$\frac{E(C_t)}{t} = \frac{1}{E(Y_i)}[c_1F(T) + c_2(1 - F(T))]$$

In this case, the components have exponential life lengths with mean θ. Hence,

$$\frac{E(C_t)}{t} = \frac{1}{E(Y_i)}\{c_1(1 - e^{-t/\theta}) + c_2(1 - (1 - e^{-t/\theta}))\}$$

$$= \frac{1}{E(Y_i)}\{c_1(1 - e^{-t/\theta}) + c_2e^{-t/\theta}\}$$

Now, to find the value of t which minimizes the expected cost per unit time, we take

$$\frac{d}{dt}\left(\frac{E(C_T)}{t}\right) = \frac{d}{dt}\frac{1}{E(Y_i)}\{c_1(1 - e^{-t/\theta}) + c_2e^{-t/\theta}\} = 0.$$

Which gives us

$$\frac{1}{E(Y_i)}\left\{\frac{c_1}{\theta}e^{-t/\theta} - \frac{c_2}{\theta}e^{-t/\theta}\right\} = 0$$

$$\Rightarrow \frac{e^{-t/\theta}}{\theta E(Y_i)}\{c_1 - c_2\} = 0.$$

Since $c_1 > c_2$ and $\theta, E(Y_i) > 0$, then $e^{-t/\theta} = 0$

$\Rightarrow T_0 \to \infty$ minimizes $\dfrac{E(C_t)}{t}$.

b. $X \sim F(\cdot)$; $Y = \min(X, T)$; $\dfrac{E(C_t)}{t} \to \dfrac{c_1F(T) + c_2\bar{F}(t)}{E(Y)}$, $c_1 > c_2 > 1$

where $E(Y) = \int_0^T \bar{F}(u)\,du$.

Now, $r(T) = \dfrac{f(T)}{\bar{F}(T)}$ and we can write $\bar{F}(u) = \exp\left[-\int_0^u r(s)\,ds\right]$.

Taking derivative of $\dfrac{E(C_t)}{t}$ and setting it to 0 yields

$$r(T)\int_0^T \bar{F}(u)\,du - F(T) = \frac{c_2}{c_1 - c_2} > 0.$$

$T = 0 \Rightarrow$ left side is 0. $r(T)$ implies left side will eventually cross $\dfrac{c_2}{c_1 - c_2}$, for some finite T.